A Reasonable Alternative to Darwin's *ORIGIN of SPECIES*

D. Bruce Fitzgerald

Captivating sequel to
"A Critical Evaluation of
ORIGIN of SPECIES"

AB ASPECT Books
www.ASPECTBooks.com

World rights reserved. This book or any portion thereof may not be copied or reproduced in any form or manner whatever, except as provided by law, without the written permission of the publisher, except by a reviewer who may quote brief passages in a review.

The author assumes full responsibility for the accuracy of all facts and quotations as cited in this book. The opinions expressed in this book are the author's personal views and interpretations, and do not necessarily reflect those of the publisher.

This book is provided with the understanding that the publisher is not engaged in giving spiritual, legal, medical, or other professional advice. If authoritative advice is needed, the reader should seek the counsel of a competent professional.

Copyright © 2017 D. Bruce Fitzgerald

Copyright © 2017 ASPECT Books, Inc.

ISBN-13: 978-1-4796-0752-5 (Paperback)

ISBN-13: 978-1-4796-0753-2 (ePub)

ISBN-13: 978-1-4796-0754-9 (Mobi)

Library of Congress Control Number: 2016920672

Scripture quotations marked KJV are taken from the King James Version.

Table of Contents

Introduction.. 5
Chapter 1 The Right Way to Read Genesis 8
Chapter 2 The World Before the Flood 20
Chapter 3 Mythological Reminiscences of the Flood 33
Chapter 4 Preparation for the Flood 43
Chapter 5 Mechanics of the Flood 51
Chapter 6 Aftermath of the Flood 57
Chapter 7 A Miscellany of So-Called Problems with the Genesis Literal Interpretation 66
Chapter 8 Conclusions 88

Introduction

Scientists today seem to be victims of information overload, as we all are, with new technical discoveries occurring every day. To ensure that this new data does not conflict with Darwin's heritage is a daunting undertaking, even if one limits himself or herself to just his own field of study. But even after such reinforcing of traditional pro-evolutionist dogma, the Theory of Evolution still falls short of explaining the major points of origin of our universe, especially in the natural history of the world's biota. What alternative can we offer?

Even after such reinforcing of traditional pro-evolutionist dogma, the Theory of Evolution still falls short of explaining the major points of origin of our universe, especially in the natural history of the world's biota. What alternative can we offer?

Let it be said that Intelligent Design, though obvious in nature, does not go far enough. It cannot explain, for instance, the fossil record without resorting to uniformitarianism (the theory of slow deposition of fossil-bearing layers of sediment) and the accompanying great epochal periods deemed necessary for the development of animals and man.

The interminably long periods needed by uniformitarianism, as we tried to show in our first book, *A Critical Evaluation of ORIGIN OF SPECIES*, are not what the fossil record shows. Creationism comes to the rescue here, answering the problem through the worldwide upheaval wrought by the great Flood, which inundated the entire earth and reshaped its topography, creating the present-day stratigraphic layers. We will bring out all the changes in the antediluvian world caused by the Flood and harmonize them with the fossil record.

> *Creationism comes to the rescue here, answering the problem through the worldwide upheaval wrought by the great Flood, which inundated the entire earth and reshaped its topography, creating the present-day stratigraphic layers. We will bring out all the changes in the antediluvian world caused by the Flood and harmonize them with the fossil record.*

But, object any who have given serious thought to the issue, this leaves us with the problem of a "young earth," an earth no more than 10,000 to 13,000 years old[1], a conclusion that scientific minds cannot accept. We answer this question by offering the correct reading of the first chapter of Genesis, which explanation does not resort to accommodation factors or straining of Scripture, but is the plain and simple communication of the Creation narrative as the author of this book believes it was intended.

In our first book we mentioned the Theistic Evolutionists, who wish to follow both the theory of evolution and the Biblical account of Creation. It is impossible to take both points of view without ending up in untenable situations. In this sequel to the previous book, we take pains to show that long-standing objections to the traditionally interpreted Creation sequence melt away under a straightforward rendering of the scriptural verses. We avoid reading anything into the story, rather accept the literal communication of each verse and its impact on the verses that follow. Any time we resort to 'spiritual' explanations of the verses in Chapter 1 of Genesis, we are guilty of interjecting human patchwork into the Divine revelation.

Any time we resort to 'spiritual' explanations of the verses in Chapter 1 of Genesis, we are guilty of interjecting human patchwork into the Divine revelation.

---

1 Bishop Ussher's famous calculation of the earth's age based on the Bible's successive genealogies was 4,004 B.C.

Chapter 1
The Right Way to Read Genesis

The first chapter of the Biblical Book of Genesis has stood as a bone of contention not for centuries, but for millennia. It has been misread, misinterpreted, and misrepresented to support the doctrine of whatever religious faction is incumbent for as long as there have been diversities of opinions about religion and the Scripture.

Because of the standoff between evolutionism and creationism, we insist above all that our interpretation from the beginning be compatible with reason, logic, common sense, sound Biblical interpretation, and verifiable in the world of nature, i.e., science.

Chapter 1 The Right Way to Read Genesis

We assert in this book that there is only one way to do justice to Genesis' first chapter—it must be based on the evidence arising from context, before and after. Because of the standoff between evolutionism and creationism, we insist above all that our interpretation from the beginning be compatible with reason, logic, common sense, sound Biblical interpretation, and verifiable in the world of nature, i.e., science.

Please, no back-door divine fiats, no unmentioned sustaining miracles, and no stretching the scriptures beyond what is literally stated. To say, for instance, that each of the six days of creation can be regarded as a thousand years in duration is only supposition and belongs with esoteric utterances of the prophets; to go even further and propose that each creation day equates to some millions-odd prehistoric epochal years amounts to theistic evolution, which we vehemently opposed in our first book.

The problem, it will be seen, is not with the scripture seeming to be 'unscientific,' but with a predisposed <u>literary</u> position. Once again, as interpreters we are always zealously molding things into our pre-suppositions: if we are leaning toward evolution, we will search to find places where some theistic/natural selection compromise can be made. If we favor creationism, we will hold hard and fast to an ultra-literal interpretation of Genesis, putting ourselves in indefensible, non-scientific positions.

Let us now put our hand to the plow with verse 1: "In the beginning, God created the heaven and the earth." Not presuming to match the erudition of a Hebrew scholar, we will nonetheless state quickly that there are three 'heavens' mentioned in scripture:

1. The sky, or earth's atmosphere (Gen. 1:20)[2]

2. The starry universe (Dt. 10:14)

3. The abode of God (Mt. 6:9)

The Hebrew word for heaven in cases (1) and (2) is *shamayim*, which shows a plural ending and therefore is indicative in Gen. 1:1 of creating both heavens (1) and (2) at this point. But we will make qualifications to this statement momentarily.

---

[2] Chapter and verse numbers in the Bible are separated by a colon: thus, Gen. 1;2 represents the first chapter, second verse of the book of Genesis.

Now the traditional interpretation of this first verse is that roughly 6,000 years ago God decided to create the human race, possibly as a counterbalance to the 'failed' experiment concerning the fall of the archangel Satan (cf. Milton's *Paradise Lost*) and those hapless angels who followed in his rebellion. If this were true, it would imply that although in the beginning heaven No. 3 was in existence, heavens Nos. 1 and 2 had yet to be created. So the traditional, long-accepted view is that God made the whole universe (less, of course, any living things) in verse 1 of chapter 1. But this view is not our best choice, as we shall see.

Let us purposely skip verse 1 for the moment and resolve it subsequently. Verse 2 is actually the beginning of 'our' creation, for it talks about the fine-tuning events that lead to the formation of a suitable home for *homo sapiens*: "And the earth was without form and void, and darkness was upon the face of the deep. And the Spirit of God moved upon the face of the waters." From a scientific point of view, we note that

1. Beyond being spherical, the earth's surface had no defined shape.

2. There were no living creatures present at all, not even a microbe.

3. The surface of the earth was shrouded in darkness.

4. Water covered the planet.[3]

What conclusions do we reach from these initial conditions?

(a) Not only was the entire universe in existence by this time, but the sun was also in place. Otherwise, what would the surface of the earth be like? It would be rock-hard, under a temperature near absolute zero (-454°F). Our atmosphere also would be completely frozen into ice crystals, as is Pluto's nitrogen and methane ice envelope today.

---

[3] The initial condition of the ocean and cloud turmoil is regarded as chaos by the early Greeks and the Semitic legends of the Middle East. There was no order, only tohu and bohu, as the Bible describes formlessness and emptiness. It leaves one with the impression that God selected the planet Earth just as he found it, in an initially unusable condition, for his grand creation.

Chapter 1 The Right Way to Read Genesis   11

(We will explain the sun's existence when the sun is mentioned again in Genesis 1:16).[4]

(b) Liquid water covered the earth to a presumed depth of 12,000 feet (just over 2 miles). This would be the depth of the oceans if the surface of today's earth were made as smooth as a billiard ball.

(c) There were no mountains or valleys, no cordilleras or troughs. Orogeny had not yet taken place.

(d) A thick water vapor canopy, resulting in 'darkness,' presumably covered the ocean, caused by the sun's unhindered evaporative action.

What would be the ambient temperature at that time? Not just warm but hot, the thick cloud cover creating an ultra-greenhouse effect. At random points, erupting volcanoes would be ejecting columns of steam high into the air, adding to the heat and darkness. Darkness and steam— along with heaving waves— nothing to commend a livable environment. So the earth in the beginning was no terra firma, but a sea of boiling whitecaps and scalding steam, a primeval formlessness.

The earth in the beginning was no terra firma, but a sea of boiling whitecaps and scalding steam, a primeval formlessness.

Following then our view of initial primordial conditions, the creation of light is simple:

---

4   Some could object that the earth may not have been quite so frigid because they say that the more abundant volcanic activity in those days would have kept the earth's surface warm. But looking in our solar system for the most volcanic bodies of all, which are Io and Europa, two of the moons of Jupiter, we see that even volcanoes on those spheres must break through heavy sheets of ice covering these heavenly bodies. Otherwise, our earth could be thought of as wearing a mantle of concrete-hard ice, crowned with a snow of crystallized atmosphere.

Verse 3: And God said, Let there be light: and there was light.

The only light source that we have talked about so far is the sun. But how did sunlight break through the thick cloud canopy? The simplest answer is that, in obedience to the command, the cloud canopy attenuated itself, or thinned out. The canopy did not go away at this stage, it was only ballooned and made diaphanous enough so that an observer standing in a boat on the earth's surface could now see the difference between darkness and light impinging on the waters.

At this point we must interject a ground rule that applies to all six days of creation: Each of the creative acts are made relative to an observer positioned on the surface of the earth. The early Hebrews (or any other ancient people, for that matter) had no concept of withdrawing themselves out into space and observing the earth from an astronaut's orbit. This preposition becomes extremely important when we discuss the 'formation' of the other heavenly bodies.

On the next day of creation, the earth continues to be groomed for human habitation. Notice that our planet was habitable to begin with (as long as we stayed in the boat!), but now we are seeing more of the fine-tuning that will bring about today's familiar world.

Verse 6: "And God said, let there be a firmament in the midst of the waters, and let it divide the waters from the waters." Before this verse 6, the cloud canopy was hanging low over the earth, so that it was impinging on the surface of the waters, yet thinned out enough so that daylight could be detected coming down from above. Now God interposes a 'firmament,' or expanse (otherwise known as our lower atmosphere) between the cloud canopy and the ocean. The clouds no longer drag the sea waves but are elevated to the heights that we know today. A comparative situation would be the thick atmosphere of the gaseous giant planets; the clouds of these planets, Jupiter through Neptune, are believed to descend until they touch the solid or liquid planetary cores, with

Chapter 1 The Right Way to Read Genesis   13

no space in between, a condition of total darkness. No firmament to the rescue for them! Yet even today we can have fog or clouds touching the ground that turn our sunny day into twilight. Notice that God calls the firmament 'Heaven,' which stretches up to the top of the highest clouds. Yet in our Biblical narrative here the canopy is still not broken up — the theory is that it will continue through the antediluvian era, until the end of the great flood, which we will henceforth call, the 'Flood.'

On the third day (verses 9–11), God makes the dry land appear. How does he do this? Simply by volcanism and plate tectonics, which we still see in operation today, on a much diminished scale. So now the seas become deeper to accommodate the rising land areas. Yet the seas were still hot, and the cloud canopy still cast its greyness over every continent (though we will find that during the Flood period and ensuing upheaval, there presumably was only one big continent to begin with.).

In anticipation of the Flood story, we must state that the land formed here in verse 9 was accompanied only by low mountains, such as the Appalachians and Urals; the 'younger' mountains, such as the Alps and the Himalayas, make their appearance at a later time.

The earth's vegetation was also formed on this day. Note that if, as we stated, the sun had not been in place from the very beginning, we would have had here an unscientific reversal, with the plant life created <u>before</u> it had sunlight from which to draw its source of photosynthesis. The appearance of vegetation in the Bible does precede, correctly, the appearance of the birds, animals, and sea life.

On the fourth day, the heavenly bodies were 'made.' But as we have previously maintained, the heavenly bodies were there all the time, just not visible to observers on the earth. The observer could tell day from night at this point, but his vision could not penetrate the ever-present cloud cover. Now, in response to divine action, the clouds are broken up and suffer further attenuation, to allow the discs of the sun and the moon to appear. This is in keeping with the wording of verses 14-18:

(v.14) "And God said, Let there be lights in the firmament of the heaven to divide the day from the night; and let them be for signs, seasons, days, and years";

(v. 16) "And God made two great lights; the greater light to rule the day, and the lesser light to rule the night."

So God made 'luminaries' for the convenience of earthlings — the light from the two pre-existing heavenly bodies now becoming discernible to anyone dwelling on earth. Though it states, almost as an afterthought, that God made the stars also, those were probably not visible until after the Flood — it was only a point of information given so that the early Hebrews would know for sure who made the stars, not some other, possibly competing, deity. Notice also that these newly visible light sources were not located in the

Chapter 1 The Right Way to Read Genesis

firmament (first heaven), as stated — they only appeared to be fixed in the firmament to an earth-bound observer.

Now would be an optimum time to go back to the beginning and discuss verse 1. Let it be affirmed for the record that it was the God of the Bible who created the sun, moon, and stars, and everything else that is visible or detectable in the universe. The only question is <u>when</u> did he do it. There are two types of 'gap' theory that try to explain this:

- One of the gap theory proponent groups says that yes, God did create the entire universe in verse 1, but that there was an interminable lapse of time before he went on to continue the work mentioned in verse 2. This would account for all of the millions (and for the requirements of evolutionists, even billions) of years to develop life upward from nothing. This concept, however, not only conflicts with the subsequent six days of creation, but is flawed in its concept of God. The Creator does not need 6 billion years of pause to think about how he is going to apply the final touches to his masterpiece.

- The second gap theory group envisions a prior creation millions of years ago in which the Devil, an early race of humans, and the fantastic dinosaurs all lived together. When Satan fell from his first estate in heaven (dragging one-third of the angels down to earth with him), the resulting conflict destroyed the dinosaurs and the first race of humans, leaving a totally 'wrecked' earth all covered with water. This is what is meant by *tohu* and *bohu*, a complete desolation.

What's wrong with this theory? Beside the fact that there is nothing in archaeological history to back this up (e.g., human and dinosaur bones are never found together in paleontological digs), the Devil and his demon minions are still at large, roaming the earth with seeming impunity, yet showing no inclination to try to wreck earth's civilizations again. And this still would not answer the "young earth"/"old earth" dilemma put forth by creationists.

Our answer is simple. As many of us know, Hebrew does not have upper- and lower-case letters; they are all one case. Further, ancient Hebrew had no periods or other punctuation marks. So a would-be translator could not know where one sentence ends and another begins — except from context. Based on that, this is how the author believes the first verse in Genesis 1 should read:

"In the beginning God created the heaven and the earth":

Notice that he only difference with this rendering versus the traditional one is the colon — making the very first verse an introductory statement as to what follows. Legitimately, we could further paraphrase verse 1 to read, "In the beginning God created the heaven[5] and the earth, and this is how he did it:".

In other words, creation really begins with verse 2. This should be obvious because if God neglected to include such great detail of labor as to how he put together the complexities of the far-flung universe from minute atomic particles to incredibly vast clusters of galaxies, holding up to possibly trillions of planets, each of which could be unique unto themselves, why would he suddenly switch gears and begin talking about miniscule modifications to a single ball of rock called the Earth? Logically, if God is the allpowerful creator, he could have created the universe in one shot with the earth finished just the way he wanted it. He could have started the Bible with verse 26 of chapter 1 and let us deduce the obvious of what went before. It is totally inconsistent for a God far superior in wisdom to any person who has ever inhabited the earth to withhold such monumental detail in creating the varied universe and then to delicately lay out only the fine-tuning steps mentioned in verses 2 through 25.

Surprisingly, the author is not alone in this conclusion. May we quote just a few of the other sources who also hold the Introductory concept as most reasonable:

---

[5] "Heaven," translated from *shamayim*, is always plural. Whether it is talking about the sky or the universe, or both, the word always appears in plural form. So in verse 1 we are only talking about the earth's atmosphere, though *shamayim* may lead us to believe that we are talking about the complete heavens.

Chapter 1 The Right Way to Read Genesis

1. Holman Bible (alternate reading)

   Gen. 1:1 "When God began to create the sky and the earth..."

   Gen. 1:14 Obviously, the lights do not <u>exist</u> in the firmament (sky), but only <u>appear</u> to come from the dome of the firmament. (4th day)

2. New Living Translation

   Gen. 1:1 (Alt.) "When God began to create the heavens and the earth..."

3. New American Bible (St. Joseph's edition)

   Gen 1:1 "In the beginning, when God created the heavens and the earth..."

4. New International Version (commentary on Gen. 1:1)

   "A summary statement introducing the six days of creative activity."

5. Jerusalem Bible (commentary)

   "The text makes use of the primitive science of its day."

6. Jewish Publication Society TANAKH (Torah/Neviim/Kethuvim)

   Gen. 1:1 "When God began to create heaven and earth..."

7. New Century Version

   "In the beginning God created the sky and the earth." (Note that "sky" would not include the starry universe.)

8. Revised Standard Version

   "In the beginning when God created the heavens and the earth..."

9. Revised Standard Version, alternate reading:
   Gen. 1:1 "When God began to create the heavens and the earth…"

10. New Revised Standard Version, Oxford University
    Gen. 1:1 "In the beginning when God created the heavens and the earth…"

11. Today's English Version, American Bible Society
    Gen. 1:1 "In the beginning, when God created the universe…"

12. New Scofield Reference Bible (commentary)
    "Genesis 1:1 … is introductory."

13. Bereishis (Genesis, a New Translation, by Rabbi Meir Zlotowitz)
    Gen. 1:1-2 "In the beginning of God's creating the heavens and the earth,
    When the earth was astonishingly empty, with darkness upon the face of the deep…"

14. The New American Bible, Saint Joseph Edition
    Gen 1:1 "In the beginning, when God created the heavens and the earth…"

15. New Century Version, Guideposts
    Gen 1:1 "In the beginning God created the sky and the earth."

16. The New English Bible, Oxford University
    Gen. 1:1 "In the beginning of creation, when God made heaven and earth…"

17. Young's Literal Translation of the Bible, by Robert Young

    Gen. 1:1 "In the beginning of God's preparing the heavens and the earth…"

18. Edgar J. Goodspeed translation of the Bible and Apocrypha

    Gen. 1:1 "When God began to create the heavens and the earth…"

19. The Catholic Bible, Rheims Douay Version

    Gen. 1:1 In the beginning God created heaven, and earth." A footnote identifies "heaven" with the "firmament." (This version has been in existence since 1609 AD.)

20. The New Oxford Annotated Bible with the Apocrypha, alternate version:

    Gen. 1:1 "When God began to create the heavens and the earth…"

21. The Living Bible (Paraphrased), Tyndale House

    Gen. 1:1 "When God began creating the heavens and the earth…"

22. The Bible in Living English, by Steven T. Byington

    Gen. 1:1 "At the beginning of God's creating the heavens and the earth…"

Chapter 2
The World Before the Flood

The antediluvian world is a tantalizing subject because so little is known about the civilization that existed for the 1,600 years before the Flood, only what is revealed in the first seven chapters of Genesis. However, there are fortunately considerably more details about the pre-Flood days than there are, for instance, for the Atlantis legend (also intriguing to speculate on, but beyond the scope of this book), except to say that Atlantis, having been destroyed by water, could be a dim pre-Flood remembrance held by the peoples of the Middle East.

The antediluvian world is a tantalizing subject because so little is known about the civilization that existed for the 1,600 years before the Flood.

Chapter 2 The World Before the Flood

What then was different between the pre-Flood world and today's world? First of all, it was hovered over by a pervasive cloud canopy — yet the canopy itself is only an assumption based on "separating the waters above from the waters below" in Chapter 1 of Genesis. After Day 6 of creation, this canopy was thick enough to hide the stars and thin enough to let the discs of the sun and moon be seen at certain times, as we suppose. We have heavy cloud covers today that prevent the sun from being seen at all and which definitely occlude the moon. So we must not suppose that the canopy was static — peradventure sometimes thick at certain localities, sometimes thin at others.

What did this canopy do for the earth? We've all heard of the greenhouse effect, which results from heat radiated from the earth's surface (originating from both the sun's incoming reflective heat and the earth's hot interior) not being able to escape into space because it is captured by the heavy cloud layer. Thus, heat comes out of the heart of the earth, but is blocked from escaping into the upper atmosphere by the vapor droplets forming the dense clouds.

Since it is generally assumed that the cloud canopy covered the whole earth in a sparsely broken spherical shell, the resultant effect would be to keep the earth a true hothouse, uniformly warm over the entire planet. Even the poles of the earth at this time could have had a tropical or subtropical climate year-round. How can we be sure of this? Simply because of the immense coal deposits found in Norway's Spitsbergen, which is above the Arctic Circle; fern imprints have also been preserved in these coal beds. Most convincing of all are the coal deposits discovered in Antarctica. The result, as many commentators have proffered, is that the earth possessed a spring-like climate uniformly over the globe.

So the whole planet at that time could be perceived as fertile— no deserts, no high mountain ranges (orogeny to be explained later). If the land area at that time were anything close to today's, the pre-Flood world could possibly have supported 10 billion people. The universal planetary warmth would have allowed all land surface areas to be fully utilized for cultivation.

There has been objection to the canopy theory because some say that the cloud cover would create too severe a hothouse effect, causing our planet's surface to become unbearably hot (a small-scale effect of what actually happened on Venus). But believe it or not, there have been scientific studies of the Canopy situation[6], and it is only at critical points of canopy height and density that the covering becomes a hot, smothering blanket. If the reader but will notice that a cloudy overcast in today's summer actually keeps the air temperature cooler than if the direct sunlight were streaming through. If we

can accept that the Genesis account of creation has any veracity at all, we can rest assured that our Deity took into account the cloud thickness and height above the earth to prime it for human habitation, just as He has done for us post-Flood dwellers.

Another curious aspect of the pre-Flood days is that there was no rain. Gen. 2:5–6 says that the Lord God had not caused it to rain upon the earth… but that there went up a mist from the earth, and watered the whole face of the ground. Presumably, the

---

[6] "Climates Before and After the Genesis Flood," Larry Vardiman, Ph.D., Institute for Creation Research, 2001.

Chapter 2 The World Before the Flood

cooler air in the mornings caused water vapor to condense out of the super-saturated atmosphere below the clouds. Fortunately for them, the antediluvian peoples were not faced with droughts or irrigation problems! And with no downpour of rain, the homes did not have to be built as sturdily as today's. Especially would the roofing be flimsier, just enough to prevent the interior from getting wetted by the morning dew. One reason archaeologists cannot identify for certain any pre-Flood dwellings is probably due to their lighter construction, thus being unable to withstand the ravages of the forthcoming Flood.

Some have supposed that the pre-Flood world had the same ideal climatic conditions as existed in the garden of Eden. Indeed, the conditions could have been the same, but no longer would all the food that the people needed grow on trees, free for the plucking: "Thorns and thistles shall it (the ground) bring forth to thee... in the sweat of thy face thou shall eat bread." (Gen. 3:17, 19) So the climate then may have matched that of Eden, but the hard toil of the earth's ground would match our agricultural struggles of today, except that we assume a condition of globe-wide fertility in the pre-Flood days at no matter what latitude one migrated to.

What about the extreme longevity of antediluvians? It wasn't just the 'righteous' people being rewarded with exceptionally long lives— everyone enjoyed it— normal old age began at 900 years! The oldest of all, Methuselah, lived to be 969. What is the scientific basis for such longevity? The only difference between our world and theirs was the brooding cloud canopy. The canopy seems to have provided protection for these early earth dwellers from the sun's ultra-violet rays. These rays today are suspected today of causing skin cancer. But the greatest damage is probably to our cells, which we cannot see. Though we welcome sunlight for our bodies to create vitamin D, the benefit could be overridden by a shortening of our life span. Notice that after the Flood, human life began to shorten dramatically. Abraham died at 175 years, Moses at 120, Joshua at 110, and after that life spans became even less than those of today (Ps. 90:10).

The Bible mentions, almost as an aside, that there were giants existing in the pre-Flood days. (These were not the offspring of fallen angels and women, as some suppose, subsequent to the first mentioning of giants). Nevertheless the offspring resulting from the angelic affair produced "men of renown" (Gen. 6:4). The Flood apparently did not eliminate the strain of giants, for they appeared again with the Philistines, especially Goliath and his brethren.

But the most significant thing that was both different and at the same time alike with our times was the universal violence. We have out-of-control lawlessness in some areas of the world today that demonizes whole countries, but back then it was every day's business, as it says in Gen. 6:11— "The earth was corrupt before God, and the earth was filled with violence." This was the cause that signaled the end of the pre-Flood civilization. We get only the barest hint of the widespread violence in Gen. 4:23: "And Lamech said unto his wives, Adah and Zillah, 'Hear my voice, ye wives of Lamech, hearken unto my speech for I have slain a man to my wounding, and a young man to my hurt. If Cain shall be avenged sevenfold, truly Lamech seventy and sevenfold'."[7] This shows the vigilante-type justice that must have prevailed before the Flood — sort of like our Wild West. Imagine what kind of world would result today if every man took it upon himself to administer 'justice' to his troublesome neighbors.

We might also ask the question, How advanced were the pre-Flood people? Genesis Chapter 4 tells us that they were artificers of bronze, which means that they had already discovered alloying copper and tin to achieve bronze, in proportions of four to one. They were also workers in iron, which means that they were knowledgeable of blast furnaces and the purification of iron ore. At this point they had already invented the flute and harp, so the aesthetic aspects of life were not overlooked. If we could go back in time and see how the three sons of Noah (Shem, Ham, and Japheth) built their houses and artifacts after the Flood, the height of the pre-Flood civilization would be revealed. These three sons obviously assisted Noah in building the Ark during the long period leading up to the Flood (120 years). If they could build something as

---

7   The fact that Lamech had two wives could be indicative of the shortage of men on the earth due to the widespread violence. This situation recurred in the centuries after the Flood, when so many men were lost at war that women had to share a husband if for nothing else but economic security.

Chapter 2 The World Before the Flood

complicated as a huge Ark, then maybe the survived human race had a good start for their new civilization. In other words, the peak of the pre-Flood civilization was represented by the knowledge and expertise of Noah and his family. So maybe we could say that the bottommost archaeological dig in the Middle East could, unknowingly to us, represent a house foundation of the pre-Flood days.

One consideration we need to make about the pre-Flood people is their language. Assuming that travel was slow in those days, with no mechanization, the people farthest from the garden of Eden expulsion site would soon be unintentionally developing their own dialects. Consider the difference in our own country over just 400 years between New England brogue and southern drawl; both parties sometimes have to strain their ears to be sure of what the other is saying. The same thing has happened between Americanese and pure British in just 250 years; if the Brit spoke fast enough, the poor American would wonder what other language the Brit was conversing in. So imagine the differences that could take place in 1600 years, between the Creation and the Flood—the people on the frontiers would probably need an interpreter to talk to the people at 'headquarters.' What does all this amount to? The greater the difference in language, the greater chance for misunderstanding—hence the babel of dialects would only contribute to short-temperedness and worldwide violence. The language barriers could not help but create circumstances of "us" and "them."

The greater the difference in language, the greater chance for misunderstanding—hence the babel of dialects would only contribute to short-temperedness and worldwide violence. The language barriers could not help but create circumstances of "us" and "them."

We need also to talk about the flora and fauna, especially the fauna. Everything was on a giant scale back in those days. Though heights of deciduous trees were on a par with what we observe today, the ferns and horsetails were gigantic. The cycads were considerably bigger than the few that are extant now. There were a number of unusual trees that did not survive the Flood but provided a fitting background for a swampy, dinosaur-frequented landscape. Everything was presumably so lush that even though Tyrannosaurus rex ruled the jungle, a lot of his prey escaped from him simply by diving into the thick brush.

Did we make a typo by saying that Tyrannosaurus rex was around in pre-Flood days? No typo— Tyrannosaurus was there from the beginning: Gen. 1:25 says that "God made the beast of the earth after his kind, and everything that creepeth on the earth after his kind: and God saw that it was good." Is there anything 'good' about a rampaging monster like Tyrannosaurus? Only that it allowed man a chance for bravado, to wipe this 'dragon' out and gain a reputation as the mighty hunter. More about this later. But there is no scientific reason why all the dinosaurs, including those winged and sea-going, shouldn't have been present on the earth from the beginning.

In the first book, evolution was shown to be indefensible when confronted by facts, statistics, and logic.
In this book, an alternative to the untenably positioned evolution is given.

But what about evolution and all the millions of years it supposedly took to develop these antiquarian reptiles? This question hopefully was answered in the first book to which this one is the sequel (entitled "A Critical Evaluation of ORIGIN OF SPECIES"). You can't have your cake and eat it too. In the first book, evolution was shown to be indefensible when confronted by facts,

Chapter 2 The World Before the Flood 27

statistics, and logic. In this book, an alternative to the untenably positioned evolution is given. The fact that the arguments in this book fit with both Holy Scripture and reason make it the "reasonable alternative" that the book title attempts to affirm.

We are not implying that all of these colossal reptiles were wiped out in the Flood. All but two of each kind were, but their lineage was carried on in the Ark. Lest we entertain objections at this point, let it be stated here that the problem of getting all the dinosaurs into the Ark was solved by admitting only juveniles. Much more about this to be brought out. At this point we see an antediluvian world populated by every species of dinosaur that paleontologists have been able to dig up today, including those that they mistake for a smaller species dinosaur when it was really a juvenile of a larger one. There was Brachiosaurus (misnamed earlier as "Brontosaurus"), the largest dinosaur that ever lived (in tonnage, not in fearsomeness) inhabiting the swamps, resorting to deep water up to the height of his long neck whenever the fearsome Tyrannosaurs would appear (however, Brachiosaurus was not the largest creature that ever lived, rather is it the blue whale, which is still with us today). Since the sauropods like Brachiosaurus inhabited the swamps as a refuge from Tyrannosaurus and his cousin variations, it follows that the theropods would linger around the swampy areas to catch a young sauropod now and then, or one who had strayed from the bounds of safety.

There were almost endless variations on Tyrannosaurus, but smaller: Allosaurus, Albertosaurus, Daspletosaurus, Velociraptor, Spinosaurus, Coeleophysis, Deinonychus, Megalosaurus, etc. The human inhabitants at that time must have regarded these monsters as 'dragons,' and probably many of them did indeed develop reputations as great hunters and heroes for their success in killing them. But at least there was plenty of green cover to run under if they missed in their first attempt. In that case only was size a disadvantage to the theropods.

But why are human remains, with almost no exceptions, not found with dinosaur bones? This question is often raised to counter the idea of a universal Flood. The poor dinosaurs, supposedly having only the brain capacity of modern-day reptiles, perished by the Flood's onslaught in the swampy areas where they lived. The humans, on the other hand, in mad desperation climbed to the highest hills and available mountains to try to escape the threatening deluge. Some few had probably managed to get into boats, but starvation or capsizing would eventually follow. When the overwhelming waters finally washed them away, their cadavers were left to float on the water's surface and hence were the last ones to be deposited when the waters finally did subside. By the time that the year-long Flood abated, what few human bones that hadn't been eaten by sharks or other maritime predators, would have been deposited on the earth's muddy surface, far above the multiple layers of slate, sandstone, limestone, silt, etc., under which the dinosaurs were entombed.

But why are human remains, with almost no exceptions, not found with dinosaur bones? This question is often raised to counter the idea of a universal Flood.

Also, in order for animal remains to be retrievable, the animal has to be relatively quick-buried. or else the scavengers and weathering would destroy every trace of the carcass. Thus, the prehistoric dinosaurs were all caught by surprise, dwelling, as we suppose, at or near the lowlands. They were dispatched by drowning and a few of them quickly buried by sediment. These ones would later leave retrievable fossils.

Prior to the Flood, then, we can summarize our assumed topographical features of the earth as follows:

- Low, worn-down mountains such as the Appalachians and Urals — nothing higher than these (to be substantiated in the post-Flood chapter).

- A lush landscape totally covered in green, like a tropical rain forest.

Chapter 2 The World Before the Flood

- No expansive oceans, but the entire earth's surface well watered, with streams, rivers, ponds, lakes, and small seas. The seas and lakes were likely deeper than our oceans of today, per Lake Baikal in Russia, which is 1 mile deep, with 6 miles of accumulated sediment below that). These reservoirs of surface water would have furnished Flood overflow when the "fountains of the great deep" were broken up.

- All of the ancient 'land bridges,' Bering Strait, Cape Horn, Australia-southeast Asia, Ceylon-India, etc. would have been connected at this time.

Were there seasons in the pre-Flood days? Obviously there were some semblance to seasons then, due to the 23 and a half degree tilt of the earth's axis. But they were moderated and kept from being extreme due to the pervasive vapor canopy. We must remember that the dinosaurs, being a type of lizard, didn't like cold weather. So it was inconceivable for the dinosaurs to have to go into hibernation to escape winter weather, because in that condition they could have been easily eliminated by earth's human inhabitants.

Additionally, the Bible mentions that God's practice was to visit Adam "in the cool of the day," showing that there were cooler and warmer diurnal periods then, as today.

In closing this chapter, we need to ask again what could have caused the demise of the formidable, seemingly invincible creatures? First, the Flood wiped out all the land-dwelling reptiles just as it did for the rest of the terrestrial animals and birds. But what about the sea-faring reptiles — would they not have survived? Which brings us to a larger question — did all of the fish survive? They obviously did, because they are here with us today. But we are also talking about the unbelievably long-bodied sea monsters, known by various names: plesiosaur, mososaur, kronosaur, elasmosaurus, nothosaurus, etc.

If all the sea-going dinosaurs managed to survive the Flood just by staying in their natural habitat, why don't we find substantially more sea-going remains than land dinosaur remains? As the

Flood made its appearance, most of the shallow-dwelling sea life was drowned, including innumerable shellfish. Those fish who were dependent on the shallow-dwelling sea life for food suffered their demise also. This left less fish in general for the large predators to feed on. But the main reason for the large-scale decimation of the remaining sea life is that the volume of habitable water had ballooned, as we suppose, at least two-fold. The amount of available fish were now spread out in a doubled volume of water, making prey hard to find. And since sea-going dinosaurs not only were at the top of the food chain, but had to stay physically at the top of the enlarged body of water to breathe, they were forced to take deeper and deeper dives for their meals; it would have been only a couple hundred feet (because of the absence of direct sunlight) before they ran into inky blackness.

In other words, food was especially hard to find on the surface because of the water's expansion into the $4\pi r^3$ volume (the mathematical formula to calculate the volume of a sphere). True to expectations, we find proportionately less of prehistoric sea-faring reptiles than land-dwelling ones. It is the author's opinion that very few of the sea-faring reptiles survived the Flood, but that the memory of them lived on in the form of sea monster legends and mythology in the minds of the Flood's eight survivors and their descendants.

A similar case occurred with the strange prehistoric mammals. We've all heard of the sabre-tooth tiger, mammoth, giant elk, and giant kangaroo, for instance. But there were other giants who engendered fear at first sight. There were the chalicotheres, 5 feet high at the shoulder, with forearms twice as long as their rear legs. Roaming the grasslands were seemingly endless variations on the theme of rhinoceros, such as the brontotheres (also called titanotheres), 6 feet high at the shoulder and as massive as a tank. The largest of all, however, was the indrocotherium, 20 feet high at the shoulder and able to eat leaves off trees the way giraffes do today. Its weight has been estimated at 15 tons, so it stood unopposed among the mammals; not even the smilodon (sabre-tooth) would threaten it (though it still did not exceed the size of the blue whale).

Chapter 2 The World Before the Flood

Why did these oversized mammals not endure after the Flood unto our times? The same reason that the ferocious dinosaurs did not: human beings and these monstrous-sized animals were sworn enemies — and the humans, though certainly more diminutive, had the advantage of brainpower, which they employed collectively, such as when the American Indians used to drive a herd of buffalo over a cliff to effect the animals' death. It is very likely that the stories of the brave dragon-killers were exaggerations of the fight against the rampaging behemoths that probably one man alone could not normally kill.

Again, it is the author's opinion that after all the animals debarked from the ark and scattered, that the humans, out of sheer terror of the memory of the old world hunting ordeals, sought to kill any size dinosaur they happened to spy; even the harmless ones, such as the duck-billed and iguanadon, would not have escaped their obsession to eliminate. On the other hand, the oversize but harmless mammals would have been killed not so much out of fear, but for meat; unfortunately, the poor indrocotherium would have had to sacrifice his whole body just so that a small community of humans could enjoy a meat meal for a few days— the majority of the indrocotherium's flesh would have lied spoiled for lack of means of preservation.

Those who wish to inject a point of skepticism at this juncture might ask, since the whales, sharks, dolphins, etc. survived the Flood, why didn't the incredibly elongated sea-faring dinosaurs also? Yet this is not a question limited to the great Flood. Even for those who still cling to a belief in evolution do we share the same problem. There have been all kinds of theories advanced as to what caused the demise of the land dinosaurs — but in addition no one to date has given a credible reason for the disappearance of the prehistoric sea monsters, much less even dare to bring up the subject. The author believes that the surface water temperature in the oceans was significantly lowered after the Flood, a condition inhospitable to reptiles; but he remains also receptive to other explanations for their disappearance.

There are many pros and cons today concerning the Flood, but an interesting sideline is the discovery in the year 2000 of remnants of human habitation along an ancient shoreline below the Black Sea.[8] This shoreline was 550 feet below a point 12 miles off the Turkish coastline. The ancient Black Sea was thought to be originally a freshwater lake that was broken through at the Bosporus by the Mediterranean circa 7,500 years ago. This 'overflowing' of course could just be another name for the subsidence of the massive Flood waters. Since then a whole string of submerged beaches has been discovered below the western shore of the Black Sea.

---

8   "Archaeologists Unearth Ancient Shipwrecks in the Black Sea," 2009, Institute for Archaeological Oceanography, University of Rhode Island.

Chapter 3
Mythological Reminiscences of the Flood

If the great Flood were recorded nowhere else but in the Biblical book of Genesis, we might have a lot more skeptical eyebrows raised than we do. But nearly all of the world's ancient traditions mention a worldwide catastrophic flood.

If the great Flood were recorded nowhere else but in the Biblical book of Genesis, we might have a lot more skeptical eyebrows raised than we do. But nearly all of the world's ancient traditions mention a worldwide catastrophic flood. Apparently the great Deluge left such a mark of fear on the survivors that the legend was passed down as a factual, indisputable occurrence from generation to generation. The message is always told somberly around the campfire, as to convey a warning, but only a very few of the versions that exist today include the conciliating rainbow, the sign guaranteeing that there will not be a repeat of the epochal Flood.

It is necessary to bring out all these renditions of the Flood story if, for nothing else, to show the uncanny agreement of detail between the source scriptures and the 'corrupted' versions that have been changed, intentionally or otherwise, by each generation as the story was passed on. Some of the versions are severely corrupted, but the nearer the version is geographically to the Middle East, the more it mimics the Biblical one.

There are universal features occurring in all the Flood traditions throughout the world:

1. A worldwide destruction of the human race, along with all other living things, by water.

2. A surviving remnant, preserved by an ark.

3. A preservation of representative animals and seeds held on the ark.

Existing today are at least 88 Flood traditions known in the world, with more being added as the languages of new aboriginal groups are deciphered. It is interesting that not all of the Flood traditions originate in the Middle East—currently known are 20 from the Middle East and Asia, 5 from Europe, 7 from Africa, 10 from Australia and the South Sea Islands, and 40 from the Americas. Some of the major legends and traditions are as follows:

Existing today are at least 88 Flood traditions known in the world, with more being added as the languages of new aboriginal groups are deciphered.

Chapter 3 Mythological Reminiscences of the Flood

The most famous of the non-Biblical Flood narratives is the Babylonian Gilgamesh epic. While digging in Nineveh, the ancient capital of Assyria, a British orientalist named George Smith came upon a huge cache of clay tablets in the library of Assurbanipal that, when pieced together, described in amazing detail an inundation like the Genesis Flood. That some exaggerations and embellishments had occurred is understandable, since the account is believed to have been written from verbal tradition and not with recourse to the Hebrew scriptures.

The story begins with an inquirer named Gilgamesh asking Utnashpistim, the hero of the Flood saga, in the same manner that Plato used in having Socrates converse with and overthrow his opponents. At this point in the story Utnashpistim had already survived the Flood and had been made immortal by the gods, though he still kept his normal physical appearance when Gilgamesh found him and desired him to recount the entire Flood story.

When the time was ripe for the destruction of the human race, the gods revealed themselves to Utnashpistim, saying,

> "Tear down thy house, build a ship!
>
> Abandon thy possessions, seek to save life!
>
> Disregard thy goods and save thy life!"

He was told to build a ship according to previously given dimensions; his ark ended up with seven stories instead of the Biblical three. Enlil, the chief god, told Utnashpistim that he would bring a "rain of misfortune" such that no one in the civilization below would suspect its durance. In this case the hero had to hide from his fellow citizens the real reason for building the ark – he told them that the reason for it being made was to transport him and his (extended) family to the netherworld, which, as their theology required, existed under the ocean depths. Apparently, Utnashpistim was rich, because he was able to hire a crew to do the work while he supervised.

He repeatedly mentions the application of pitch and asphalt, as if to prepare for a long flotation. However, when it came time to enter the ark (with all the animals), Utnashpistim did not have a deity to shut and seal the door behind him, as in the Biblical account – he had to close it somehow, keeping the leaks out, himself!

In the Gilgamesh epic, the rain did not arrive in a steady downpour, but in a series of violent storms, which also flooded their civilization's existing dikes and forced them to give way. In this case the incessant rain lasted for only six days and ceased entirely on the seventh. Gilgamesh acknowledges that every human being had been wiped out except those on the ark, which came to rest on "Mount Nisi" after only one day of calm.

Here the resemblance of the Gilgamesh epic with Genesis is bizarre: After six days of the Flood's subsidence, Utnashpistim sends out a dove to search for land anywhere. The dove returns, finding only the original ark for a resting place. Then Utnashpistim sends forth a swallow, but she eventually returns also. Finally, a raven was released, and it did not return.

Just like Noah, Utnashpistim offers a great sacrifice, but does not shed animal blood as Noah did. As a reward for Utnashpistim's saving of the human race, the chief Babylonian warrior God, Enlil, made him and his wife gods also, but they were limited to dwell on the earth, becoming guardians at the mouth of great rivers.

The similarity between the Gilgamesh and Genesis accounts requires a derivation from one source. There are some who say that the Genesis Flood story was taken from the Gilgamesh one; but the Genesis account is reasonable to the last detail, as we will show in Chapters 4, 5, and 6, whereas the Gilgamesh epic is fraught with exaggerations and uninhibited embellishments. The Gilgamesh version assumes multiple gods that quarrel among themselves (as do those of the Greeks); its ark had seven stories, making the whole vessel cube-shaped and unseaworthy; each side of the cube was 120 cubits (180 feet)! Enlil, the god who unleashed the Flood, was angered that some humans had dared to survive

Chapter 3 Mythological Reminiscences of the Flood

the Flood, whereas the Genesis account shows that a surviving remnant was strictly God's intention.

Not all the Flood legends, but the Gilgamesh epic and the Bible mention a 'sign' that such a flood would never occur again; the Bible establishes the rainbow and the Gilgamesh epic the jewels of Ishtar (the chief Babylonian goddess).

Certain scholars, including some Christian ones, attribute the Hebrew Flood account to Babylonian sources, the latter of which they say came first. Other scholars say that both Flood accounts originated from a single, earlier source. The author favors the majority viewpoint, that the Babylonian account was handed down verbally by tradition going all the way back to the actual Flood, whereas the Hebrew account was part of later revelation given from above. Actually, there are several different Deluge versions originating from the Tigris-Euphrates area, but not one of them is as realistic and informative as the Gilgamesh epic. Once again, it is evident that the universal watery catastrophe made such a deep impression in the minds and memories of its few survivors that the story was passed down from generation to generation with feelings of awe and great respect, not wishing to provoke the gods into anything like that again.

There are two other Assyrian traditions of the Flood. In the second tradition after Gilgamesh, Atrahasis ("the exceeding wise") complains to the gods that he has no experience in building a ship (in contrast to Utnashpistim), and he is forced to ask the god Ea to make a design for him of a floatable vessel.

The third Assyrian tradition does not mention a hero but focuses on the fluctuating moral condition of their society. The gods visit their disfavor on the Ninevites first by a universal famine; when the population repents (note the similarity to the book of Jonah), the famine ceases. But not long after their reprieve, the Assyrians drifted back into their old sinful behavior. This brings on a pestilence and barrenness in their flocks and fields. Finally, after the immoral behavior had become worldwide, this was too much for the gods, who then swept away the entire Mesopotamian population, save the Ark voyagers, in the great Flood.

Two Babylonian Versions

In addition to the Assyrian stories, there were two Babylonian versions of the Flood, written almost contemporaneously with Gilgamesh. In the earlier of the two versions, the hero (whose name was lost from a fragmented clay tablet) was saved in a ship which also floated out the storm after seven days. When the ark was grounded, the hero opened the ship's door, and the sun god Utu appeared; the hero sacrificed an ox and a sheep and bowed before Anu and Enlil to receive the gift of immortality. This represents a variation of the Gilgamesh epic.

In the second Babylonian tradition, the hero's name is given as Atramhasis (vs. Atrahasis in the Assyrian version). After the moral failure of the human race at that time, the god Enlil sends a great flood-storm accentuated by towering clouds and strong wind. Enlil warns Atramhasis that he must leave behind his earthly possessions if he wants to survive the Flood. It is possible that the two Babylonian versions and the Gilgamesh epic are variations of the same legend.

Two Sumerian Versions

Historically, the Sumerians pre-dated the Babylonians, though both had similar cultures, both using the cuneiform text. The first tradition, quite similar to the Babylonian ones, shows the priest Ziusudra, as he just finished fashioning an idol of wood to worship and anticipating it to be an oracle from which the gods would speak to him. The priest was saved in a ship that weathered the Flood for seven days. When he opened the ship's covering, there (again) was the god Utu. Ziusudra sacrificed an ox and a sheep, doing obeisance before Utu, and then receiving the gift of immortality (but still, as with Utnashpistim) being confined to the geography of earth.

An interesting quotation appears in the ancient Sumerian List of Kings, all of whom are assigned fantastically long periods of reign: "(During the antediluvian era), there were five cities: Eridu, Badtibira, Larak, Sippur, and Shuruppak; eight kings ruled over them for a combined total of 241,000 years. (But then) the Flood

Chapter 3 Mythological Reminiscences of the Flood

swept over the earth." After the Flood, the cuneiform records say that when the gods allowed kingship to be restored, the first kingdom was at Kish.

A third version is given in the work *Babyloniaca*, written in Greek by Berossus, a priest of Marduk, in 275 BC. The manuscript of his writing has been lost, but so many Greek and Jewish writers have quoted from it, that we are able to piece together at least the Flood episode from *Babyloniaca*. If these quotes are correct, there were a series of antediluvian kings, all of whom were supposed to have had an abnormally long life, but the tenth, Xisuthros, was warned by the gods to prepare a ship to save himself, his family, and friends. But after disembarking from the Flood on a mountain in Armenia, only he, his wife and daughter, and the ship's pilot were taken from among the survivors to be immortal with the gods.

Various Flood Legends of the World

- The Greek legend of Deucalion mimics closely the Biblical account, adding that Zeus inspired the animals in the ark with "reciprocal amity," so that they would not devour each other.

- According to the Egyptian historian Manetho (ca 250 BC), the ancient Egyptians mixed together the tradition of the Flood and the commemoration of the dead: The priest placed an image of Osiris (god of the Underworld) into a sacred ark and launched it out to sea, watching it until it disappeared from sight. The ceremony was observed every 17th of Athyr, strangely corresponding to the Hebrew Flood date of Cheshvan 17th.

- Ovid, the Roman poet, writes in his *Metamorphoses* that Jupiter could no longer bear looking down on the wickedness of man and decided to rain down "incessant showers." Neptune joined in and gushed out all the springs into fountains swelling every stream until they overflowed. The two who escaped in a barque were Deucalion, a pious and devout man, and his wife Pyrrha. As the Flood waters gradually subsided, they beheld the land literally "lift up" to lower the water levels. Ultimately Deucalion and Pyrrha repopulated the earth by throwing stones behind themselves over their shoulders, wherein those thrown by Deucalion changed into men, and those thrown by Pyrrha into women.

- It is interesting to note that, in the Hindu version of the great Flood, Manu (the equivalent of Noah) saves himself and eight people aboard his ark, just one more than were saved in the Hebrew version.

- Also, in the Satyavrata (one of the Hindu holy books), after the ark touched dry ground, Manu became drunk with too much "mead" and was discovered unclothed by his three sons, two of whom quickly covered him with a sheet; a curse fell on the third son, who witnessed his father naked but did not cover him.

- Brahma (not the god Brahman) was born on a lotus growing from the navel of Vishnu during the universal Flood.

- Chinese traditions show Fah-he escaping from the Deluge with his wife, his three sons and their wives, also making a total of exactly eight.

- Plato in *Timaeus and Critias*, speaks of Atlantis, the sunken city that had grown to a great population but had been overcome by wickedness and therefore was given up by God to be deluged.

- The Quiche Indians of Canada say that "At first all was sea — no man, animal, bird, or green herb — there was nothing to be seen but the sea and the heavens.

Chapter 3   Mythological Reminiscences of the Flood   41

- The Cholula Indians of Central America had an amazing account of the Tower of Babel, a structure built not long after mankind's recovery from the Flood.

Universality of the Flood Account

Peoples who have Flood stories contained in their mythology which are a variation or obvious corruption of the Biblical account are worldwide, on all continents:

Sumerian	Burmese	African	Philippines
Babylonian	American Indians	Tamil (Sri Lanka)	Borneo
Persian	South American Indians	Mongolia	Fiji
Egyptian	Mexican Indians	Greenland eskimos	Palau
Arabian	West Indies	Kamchatka (far east Russia)	Australia
Greek	Alaskan Indians	Formosa	New Zealand
Chinese	Lithuanian	Malay peninsula	New Guinea
(east) Indian	Transylvanian gypsies	Indonesia	Melanesia/ Micronesia

As earlier stated, there are at least 88 Flood traditions, of these 40 are from the Americas. Thus, the western hemisphere, from Tierra del Fuego all the way up to Alaska, have the highest representation of peoples with strong recurrences of the great Flood.

In summary, we can re-emphasize the concurrent themes in all the Flood stories:

- The highest deity in each culture sends a retributive flood that destroys the entire human race and all other living things (except, of course, the sea-dwelling creatures — to which we spoke more on in Chapter 2).

- Those deemed worthy of being spared by the deity (always an extremely few in number) build an ark as a means of escape.

- A seed of all the animals is preserved to perpetuate the continuance of animal life. The fact that the animals would have to be fed for a duration after the ark touched land implies that the ark-builder remembered to take various grains aboard also.

- The reason for the Flood (though not occurring in 100 percent of the Flood traditions) is that mankind's persistent wickedness had made him vile in the eyes of the deity.

Chapter 4
Preparation for the Flood

The Bible doesn't say, but Noah was warned of the coming Flood in sufficient time to build the ark before God's deadline. The ark would turn out to be an ocean-liner-sized vessel. He also had the help of his three sons, Shem, Ham, and Japheth. Since Noah was also a patriarch descended from the Godly line which included Adam, Enoch, and Methuselah, it is possible that he was wealthy enough to employ additional hired help for the massive project.

Fortunately, Noah received blueprints for the ark directly from above, even if they were only verbally given, just as Moses received details from above of how to build the Tabernacle in the wilderness (Heb. 8:5). For every part of the ark he was to use gopher wood; since this word occurs only once in the Bible (Gen. 6:14), there is no way to compare its usage contextually elsewhere. The best guess is the *cypress*, which is a resinous wood; this helped to keep the boards of the ark from rotting but still did not provide the extensive supply of pitch needed. This pitch would not have been the thick, black tar that results when heavily matted vegetation is compressed under layers of sediment and heat.

Before the time of the Flood, the source for pitch was pine trees. Surprisingly, even in our times, in the days of wooden ships, pine pitch was one of Sweden's chief exports; the commodity reached a peak in 1863, when Sweden exported 227,000 barrels of pitch, all derived from the wood and roots of their native pine. So Noah would have tasked his servants to gather large quantities of pitch from a presumably nearby pine forest. As a precaution, he was told to pitch the ark from "within and without."

God gave Noah the exact dimensions of the ark, which were to be:

Length: 300 cubits (450 ft)

Width: 50 cubits (75 ft)

Height: 30 cubits (45 ft)

The cubit is roughly 18 inches (from the elbow to the tip of the large finger). But in Biblical times there was also the "great cubit," which was 20.5 inches. If we used this length, the ark would have been significantly greater in size.

The vessel was to have three stories; the top story was to have a window 18 inches below the very roof of the vessel. No other opening was specified, except for the side entrance door at the bottom. Which brings up the question: How did Noah keep the door, that would be submerged during the Flood, watertight? — we will get to that.

The ark was essentially a flat-bottom barge, virtually immune to capsizing. Scholars who have studied the ark's dimensions say that it has

Deck area 95,700 square feet (three decks)

1,396,000 ft^3 of volume

displacement of 43K tons (1 ton = 100 ft^3 of usable storage space)

The ark's displacement was equivalent to that of the Titanic's.

Chapter 4 Preparation for the Flood

In spite of these spacious dimensions, some have waxed skeptical that pairs of all the world's animals could fit into the ark. First of all, in response, it is estimated that the ark had the capacity to hold the cargo of 520 railroad boxcars. In overall size, the ark was equivalent to a World War II aircraft carrier.

Noah's boat was intended to be a safe repository for two of each animal species of the world, which, according to Ernst Mayer, a crown prince of evolutionism, amounts to 1,000,000 species. However, since this figure is dated, modern biologists are thinking big and estimate up to 14M species.[9] The insects of this number would easily fit into the ark — but what about the larger, bulkier mammals? In fact, what about the dinosaurs, of which 250 species (but only 100 genera) have been identified? The answer, as we can reasonably assume, is that Noah took aboard only juveniles. How did he accomplish this? He took what God guided to him, in appropriate pairs.

The next question typically raised is how did Noah feed all these animals? The answer is, that he didn't have to. The food brought aboard along with the animals (Gen. 6:21) was meant for a different purpose. The only reasonable explanation is that all the animals were induced into a coma-like hibernation for the year that they were confined to the ark. How can we propose something like this? Because many animals today in their natural state go into suspended animation for a season: The bear (though not into a really deep hibernation), arctic ground squirrel, box turtle, dor beetle, garter snake, bat, marmot, chipmunk, etc.

In the heat of summer, several reptiles, notably selected frog species, go into what is called *estivation*, a stupor similar to hibernation. But if the animals went into a somnolent state for a year, why was Noah commanded to take food for the animals? The food was not meant to sustain the animals during the Flood, but to keep them alive after the ark had landed. The hills would be bare or just sparsely covered with soggy vegetation not worth grazing upon. Furthermore, assuming that vegetation only began to grow on the exposed land say, a month after the waters began to recede,

---

9   One definition of *species*: A group of organisms that can interbreed, producing fertile offspring, and who do not mate with other species.

why didn't the carnivorous animals (lion, wolf, etc.) tear into and devour all the herbivorous ones?

The answer might be that since God brought in seven pairs of all the clean (sheep, goats, deer, etc.) instead of just two, perhaps they provided initial food for the meat-eaters. More likely, God gave them a temporary period of vegetative ingestion, as He had done in the peaceable times in the garden of Eden. Besides, as in the case of the dinosaurs, the meat-eaters were probably all juveniles and would not have developed into full savagery. Note also that the great majority of evolutionist biologists today consider that the earth's land area at that time was formed by one large supercontinent with seas all around it, so that animal pairs would have had no trouble wending their way to the ark.

Did Noah face any resistance from his neighbors while constructing the ark? It is pure conjecture as to what threats he may have received, though the Bible states that the human race had at that time degenerated to a state of universal violence. Of one thing we can easily assume— that Noah was the object of mockery by his erstwhile friends and neighbors, and as word spread far afield of what he was doing, possibly ridicule from the whole planet. Today we could relieve some of the pressure by saying that we were building a theme park which we would charge money to see. But Noah, a "preacher of righteousness,"[10] would not have tried to evade the issue — he would have forthrightly warned his countrymen of the ark's purpose.

Thus the Lord "shut him in," which was the second of the miraculous events accompanying the Flood (the first being the rounding up of the male/female pairs) — for the door of the ark required sealing "from without and within" to keep the vessel watertight. Those who try to explain the Flood's events purely from a naturalistic viewpoint will bog down several times in this story in attempting to rationalize the events of Genesis Chapters 7 and 8.

As far as speculation about other human beings surviving the Flood besides Noah and his family, one might ask, Did they not have boats back in those days? Obviously they did, but probably

---

10 II Peter 2:5

Chapter 4 Preparation for the Flood

not large sea-going vessels. As we will postulate later, the water level in antediluvian days was significantly lower than it is today. We mentioned the Black Sea case, where the water level was 550 feet lower in ancient times.

In frantic desperation as they saw the end coming, these antediluvians no doubt jumped into their existing boats, even into the largest of them. But how could they survive in such crafts for a year without food? Even if they had time to hastily stock sustenance aboard their boats, it would never last even a fraction of a year. Worse than that, as we shall show in the next Chapter, the undersea volcanoes which were broken up and called into service at that time would have generated huge waves, even tsunamis that would engulf and summarily capsize whatever boats had been utilized at that time.

Did Noah himself know how wicked the world had become? Jesus prophesied of these times, saying, "As the days of Noah were, so shall also be the coming of the Son of man. For as in the days before the Flood, they were eating and drinking, marrying and giving in marriage, until the day that Noah entered into the ark."[11] Thus, if Jesus spent no words of remorse for the demise of the antediluvian civilization, we can confidently assume that these people, down to the last man and woman, had gone too far into iniquity to be redeemed. So building a huge ark may not have seemed so far-fetched in Noah's mind as he witnessed day after day the hopelessness of his close and distant neighbors.

What about the strange spectacle that the antediluvians watched as the animals marched in two by two? They might have gotten a chill down their spine and a sense of foreboding judgment. (The entrance door must have been larger than most of us realize, if two baby elephants and two giraffes had to fit through.) All of Noah's neighbors sensed that something unheard of was about to begin, yet they still did not grasp the importance of the ark. None of them rushed madly to the ark's door, exclaiming, "Noah, save us too!" Wouldn't any of us be filled with awe at the sight of pairs of animals pouring into the ark for seven days? Yet the fact that the earth dwellers had gotten away with their sinful

---

11 Mt. 23:37-38.

living for so long, experiencing no immediate retribution, jaded them into thinking that nothing extraordinary was about to burst forth and suddenly end their lives. The Bible doesn't say so, but if just one of Noah's acquaintances had suspected that judgment was coming, he could have barged into the ark while the animals were entering and thereby ensured his salvation.

Another question that usually arises is, since there may have been several billion people on the earth at the time of the Flood, why do we not unearth even a trace of their fossils? Because human beings then in their frantic desperation to survive would have sought the highest ground as they fled the rising waters; they and the birds would have been the last to succumb — hence their remains, being exposed on the waters' surface would have rapidly deteriorated instead of being preserved under layers of sediment. Their bloated corpses would have provided the sharks a royal feast for a whole year.

Did Noah have any personal doubts as to what he was doing as the local buffoon, laboring unceasingly for 120 years? God did not tell Noah where the water would come from to wipe out mankind of his day. He did not tell Noah that the Flood would begin with rain (as the antediluvian people had never seem rain before, only the heavy morning dew), and notice was not given until seven days before the Flood began.[12] Thus the Lord "shut him in."

Back to the food issue: Noah would have made a large granary for threshed wheat, though they wouldn't be able to bake bread. Then he would have had to take on innumerable earthen vessels to contain the olive oil needed to go with whatever bread and vegetables they were able to keep from molding. To drink, they would have required innumerable wineskins, full of wine and/or possibly fresh water. They would have been sure to bring seeds aplenty aboard, though eventually the seeds being exposed as the waters subsided would have refruited the earth.

Did Noah and his family get tired of the same diet day after day for a year? But probably Noah had also stored nuts, honey, raisins, and other fruit that would not spoil. Living as a vegetarian for a year was a small price to pay for escaping with one's

---

[12] Gen. 7:4

Chapter 4 Preparation for the Flood   49

life. Noah probably had the foresight to bring an abundant supply of dried fish with the ark. Besides, it is likely toward the end of the Flood that they were able to fish protein out of the subsiding waters, assuming that Noah had the foresight to bring along fishing gear. So all in all, Noah and his family would have had to store enough edibles to last for 2 years.

Was Noah in danger of coming down with *scurvy*, as had so many seamen have in our post-Flood history did? Not if God advised him to bring along onions in addition to dried fruit. Since it would be too dangerous to have a fire aboard for cooking, Noah and his family would have had to get along with eating things raw for a year — but raw vegetables actually give them better nutrition. Yet what a thanksgiving they must have offered to God when after a year of their debarkation they were able to gather their first harvest from the earth!

With the whole community probably calling him " Noah the *non compos mentis*," how did he make a living for 120 years while he devoted most of his time to building the ark? Most likely, while he and his three sons hammered away at the ark, he put his wife and three daughters-in-law busy with cultivating the field: they would have needed to gather copious quantities of grain to take aboard the ark and sufficient vegetables and staples to feed themselves up to the time of departure. (Those today who are prone to cabin fever can just imagine how it must have been to be cooped up in a single ocean-going vessel for more than a year.) But at least Noah did not experience any work stoppages due to rainy weather.

What about water? The people on the ark could not drink the sea water because of its salt content, though because of the vertical expansion of the seas, it was probably less salty than the oceans of today. Could Noah catch rainwater from the ark? Since we are told that there was only one window raised above the ark's roof and the rest of the ark was apparently covered by a water-tight roof, there would be limited human access to the outside. Besides, with the continual heaving waves and tsunamis, a human being would very easily be swept off the roof.

Considering the total aftermath of the Flood, especially the tremendous depths of sediment that were deposited at all possible places on the pre-Flood world, one would quickly realize that the Flood was a great obliterator, wiping clean every trace of human occupation, in spite of our wish today of finding even a few artifacts; every trace of human occupation would have been swept clean.

Take, for example, the famous White Cliffs of Dover. This cliff, 350 feet high from the English Channel water surface, consists of white chalk, which in turn is the skeletal remains of single-cell planktonic algae. These skeletal remains make up the entire 350-foot height of the 10-mile long cliffs. Contrary to geologists' claim that the chalk was formed by ice-age glaciers pushing southward from the arctic circle, they are much more reasonably explained by diluvial deposits. Imagine undertaking to dig down 350 feet in hopes of finding what the antediluvian land surface actually looked like.

Chapter 5
Mechanics of the Flood

According to the Bible, the first appearance of rain lasted 40 days. Where did this rain come from? We had mentioned earlier "the waters above the firmament," which we interpreted to be a vast cloud canopy full of water vapor, but with no dust particles to condense on and form a nucleus for rain. How thick would the cloud canopy have to have been for it to rain 40 days?

With no dust particles to condense on and form a nucleus for rain, how thick would the cloud canopy have to have been for it to rain 40 days?

It is estimated that if all of todays' visible clouds were condensed and emptied, they would only produce 1 inch of rain worldwide. Further, based on a rainfall of 0.5 inch per hour for 40 days (a substantial rate), when the vapor canopy fully emptied itself, it would only have deposited a water column of 34 feet worldwide, on land and sea. For those who are doubtful about this much water emptying from the heavens, consider that the rainfall rate could have been reduced to a fine spray (like a "farmer's rain") for 40 days without violating our concept of canopy sufficiency. This would certainly drown the lowlands, but what about the high hills and mountains? The Bible said that the Flood rose to a height of 15 cubits (approximately 22 feet) above the highest mountains (Gen. 7:20). Our answer to this dilemma follows.

The average ocean depth on today's planet is 12,000 feet. If all of the highest land elevations were planed down to a smooth, average level, the universal ocean that would result would cover the earth to a depth of 8,000 feet (1.5 miles), with no land showing at all. So from whence did all the inundating water originate? In addition to the "windows of heaven" being opened in Gen. 7:11, the same verse tells us that "the fountains of the great deep were broken up." What are these fountains?

The only reasonable explanation is that they were undersea volcanos, which spewed forth columns of smoke and steam far above the water's surface. There is plenty of geologic evidence for this today, as when in 1962 the 1.5-mile island of Surtsey was formed just off the coast of Iceland from towering steam clouds and hot ash. Yet in Noah's case, there were not just a few volcanoes activated, but hundreds and perhaps thousands of them, the net effect of such volcanic activity being as shown in Figure 2.

But one might object that such activity could never cover the highest mountain peaks with water, those such as the Himalayas or Alps. Our answer, which cannot be proved at this time, is that all of the 'younger' mountains, that is, the Rockies/Andes, Alps, Caucasus, and Himalayas did not exist before the Flood. These will be explained later. The antediluvian world is believed to have had only low-lying, 'older' ranges such as the Appalachians and Urals and possibly some no-longer extant ranges. The conjectured cross-section of the early Earth's surface is shown in Figure 1.

Be it noted that the volcanic activity did not occur only at the Flood's beginning, but most likely throughout the whole Flood period (371 days). One can hardly imagine what a turbulent sea

Chapter 5 Mechanics of the Flood   53

Noah had to deal with. Consider also that 70 percent of the ejecta of volcanoes was water. The steam thrown up by thousands of volcanoes, all active at once, condensed into rain; the process could have gone on for 40 days. The innumerable volcanoes provided more than sufficient fine ash for nuclei of water droplets.

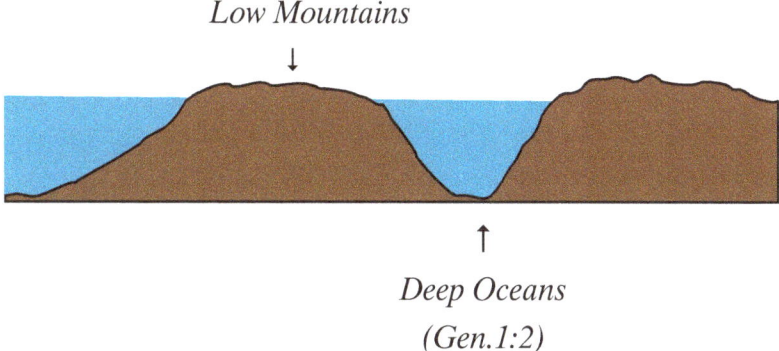

Figure 1. Typical Cross-Section of Pre-Flood Topography

Figure 2. Overflowing of Existing Land through Breakup of the "fountains of the great deep" (Gen. 7:11)

Figure 3. Continued Rise of the "fountains of the great deep" by the End of the Flood, forming the 'young mountain' ranges (Subsidence of Flood Waters Resulting in Greater Surface Water Area than Land) (Ps. 104:6–8)

So how did Noah and his family fare during their year-long siege on the ocean-covered earth? Fortunately, the animals did not need feeding, assuming that they were in some sort of suspended animation, wherein also there would be no need to clean up animal waste. There was only one window (Gen. 6:16), located at the very top of the ark (see Figure 4), a laborious place to bring waste matter up from three stories below, if it were not for hibernation.

What was there to do then? Nothing but stand in awe at what was happening, transpiring before them as if in a dream. Then the problem also comes up: if there were only one window at the very top of the ark, how was Noah able to see in the lower levels? The answer would be the creation of a large vertical shaft built during construction from the window down to the lowest level. Assuredly, the ship design diverted the splashing of the waves against the ark so that they did not overflow into the shaft.

What about boredom? With no lights in the ark, the nights must have been intolerably long. And when night was over, there was only the single shaft of light coming down from the window — certainly a time of mutual suffering for Noah and his family. And whenever giant waves were able to beat against the window, sending spray down the shaft, Noah must have wondered if that were a sign of their premature end.

When the highest mountains were covered and the draft of the ark pulled no more than 15 cubits, things may have settled down. But not really. It was time for the young (as all geologists will agree) mountains to be formed. Yet we reach one of the greatest problems of the Flood epic — where did the receding water go?

Certainly geologic evidence shows that in times past the oceans were shallower; so much of the receding water did not go anywhere, but stood in place as a deeper ocean. It took only 40 days for the Flood waters to reach their maximum height, but it took 221 days for the waters to recede so that Noah could leave the ark. Therefore, the subterranean upheavals and further subsidence of basins must have been continuous rather than a one-time event, because Gen. 8:2 says that after 150 days, the waters were assuaged.

Chapter 5 Mechanics of the Flood   55

Some of the water, of course, evaporated, and began forming what we observe as present-day clouds; but as mentioned earlier, condensation of all the world's clouds at once would add no more than an inch of rain uniformly to the earth's surface, But a great volume of the water did not go anywhere — it stayed put, as previously stated, accounting for the much greater depths of the oceans and the fact that they now covered 74 percent of the earth's surface.

The question remains: Where did the excess water go? We can answer this by going back to Gen. 1:9: Working with an earth covered with water (depth unknown, but likely no more than 1.5 miles), God raised continents out of the water, giving at the same time, as we presume, abyssal depths for the water to collect in. Thus, God lifted the earth in some places and sank broad undersea basins in others. Thus there was no diminishing of volume of water.

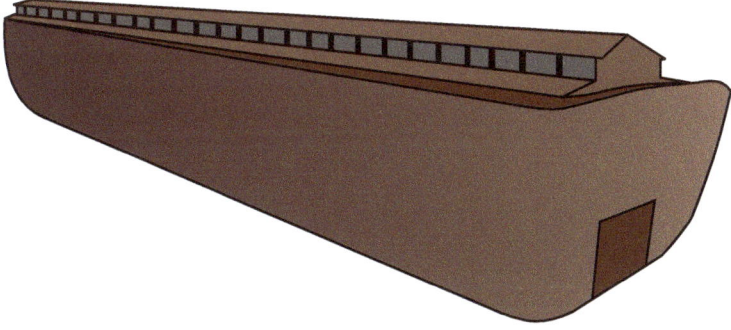

Figure 4. One Conception of the Ark

In Noah's case He did the same thing: It was at this time that the younger mountains were raised (Rockies-Andes, Alps, Himalayas, etc.) and that the sea floor again suffered a radical collapsing. The same mechanics as were used in Gen. 1:9 could have been employed here. So the gradual recess of the waters from Noah's mountain was (as we presume) due to the slow but steady decline of subterranean basins and trenches. The only difference between Noah's experience and that of Gen. 1:9 was that Gen. 1:9 was effectively instantaneous (being accomplished within one 24-hour day), whereas Noah's view of land appearing would have been painfully gradual. Psalm 104:6–9 implies that at the end of the Flood, God raised up the mountains and sunk the valleys correspondingly. Some of the Flood waters obviously became frozen and were allayed into the great ice-age glaciers that shielded North America and Europe for possibly centuries.

Chapter 6
Aftermath of the Flood

The earth was covered with water for just over a year, though the ark came to rest on the "mountains of Ararat" after 5 months. (Even though the rains had stopped after 40 days, the waters continued to rise for another 110 days, presumably due to worldwide volcanic action pushing up from the abyssal deep.)

The presence of the 'young' mountains has always puzzled geologists. After all, every young range (Himalayas, Alps, Rockies, Andes) are abundantly covered with marine fossils at their tops. This could only mean that the highest mountains were once under water.

Which presumption brings us to the crux of the matter: The presence of the 'young' mountains has always puzzled geologists. After all, every young range (Himalayas, Alps, Rockies, Andes) are abundantly covered with marine fossils at their tops. This could only mean that the highest mountains were once under water.

For the Biblical Flood to have reached an elevation and volume that would cover these great mountains, would have been impossible, considering that the Flood water depths would have reached to 5 miles above the earth (the height of Mt. Everest), an impossible abyssal volume both before and during the Flood.

Consider that before the Flood, the earth only had low, worn-down mountains such as the Appalachians and Urals, exceeding no more than 6,000 or 7,000 feet in elevation. On what occasion then, did the 'young' mountains originate? The answer is, once again, from the Bible's mentioning of the "fountains of the great deep." At the very same day that the torrential rains began, the colliding continental plates started breaking up the ocean floor.

This crustal upheaval did not cease when the rains did (after 40 days), but apparently continued for the full year of the inundation. The solution to the origin of the young mountains is that they were pushed up during the final period of the Flood, creating also a corresponding sinking of the ocean floor and providing great basins to take in the surging benthic rivers and currents. The rise of the great peaks forming today's highest mountain ranges would have left substantial collapsing floors under the ocean; these mountain peaks did not appear to Noah until the end of the 371 days, showing that the young mountains rose near the very end of the Flood period. The extreme ocean depths such as the Marianas Trench (7 miles) were probably formed in reaction to the rising new peaks.

What now are the differences in climatology brought about by the Flood? Before the Flood, in the steady-state antediluvian world, we have supposed the following:

- Unbroken cloud canopy overhanging the whole earth
- No seasons
- Low, ice-free mountains

Chapter 6 Aftermath of the Flood 59

- No atmospheric storms
- Rainless sky
- Uniform, warm humid climate

As an interesting note on the mildness of the pre-Flood climate, Alaska once supported mangroves, palm trees, nutmeg trees, and macasser oil trees. Also, it is estimated that Alaska and Siberia supported up to 5 million mammoths before the Flood. Sequoia tree fossils are found all the way north through Canada to inside the arctic circle.

In contrast to the climate before the Flood, the post-Flood climate (which may have taken several centuries to stabilize) showed that

- Rainbows first appeared
- The difference in seasons could be felt
- The poles froze over, permanently
- Wind began to play a major role in the earth's weather
- Coal, formed from compacted vegetation, has been found in Spitzbergen and Antarctica. (It takes 10 to 14 feet of vegetation to produce a foot-thick seam of coal.)
- Electrical storms appeared
- For a time, what are now deserts bloomed like a rose. (Plant fossils, along with man-made implements, show that the entire Sahara was once inhabited; likewise for the Gobi.)
- Snow was noticed for the first time, becoming a permanent cap on many mountains
- Agricultural seasons stabilized — no longer could food be grown at just any time of the year

- Predictable weather patterns emerged, such as wind in March, snow and ice in winter, hurricanes and monsoons in the fall.

Interestingly, seamounts, located 1,000 fathoms below the earth's surface, are flat-topped, showing that they had pre-Flood origins and not volcanic ones, but also show fossil remains. Since volcanic rocks are found worldwide (including at the bottom of the oceans embedded within sedimentary rock), volcanoes must have been active during the whole period of the Flood.

What are some of the problems that Noah had to deal with after everyone, including the animals, debarked?

- The food stored on the ark had to be distributed to the animals, taking into account each species' special diet.

- Noah had to plant seed quickly, hoping for a quick harvest

- Fortunately, he was able to use the ark as a shelter from the now unpredictable rains

- Hides had to be harvested from the 'clean' animals to enable the humans to withstand the cold mountain temperatures

- Once they discovered that it was warmer at lower altitudes, they would begin building permanent shelters; it may have meant cannibaliizing the ark.

What happened to the ark, or at least to its remains? Noah's three sons were probably tasked with hiking back and forth up the mountain to get needed things, but after enough treks up the long, arduous mountainside (17,750 feet), they probably stopped the practice because they could carry only a backpack-full or a donkey load at a time. Once the mountain became covered with winter snows, they probably had an impossible time retracing their steps to the ark.

Did the wood with which the ark was built rot? Not if gopher wood was basically resinous, especially if pitch covered the outer surface of the ark liberally. But 3,500 years is a long time. Our only

Chapter 6 Aftermath of the Flood

hope is if the ark, after endless snowy winters, had become crusted over with ice, like a glacier, could we reasonably expect to uncover it someday. (But the Bible says that the ark landed on the <u>mountains</u> of Ararat, not necessarily on today's Mt. Ararat (Gen. 8:4).

From the time that the Flood began until when Noah stepped out of the ark was 371 days. After they arrived on dry land, the flood waters remained on the earth for nearly a year without much change. Even as Noah and his family set up camp and began to sow seed, the waters continued to subside very slowly, continuing in some areas for years, or even centuries.

On Pico Alto in the Azores (17,600 ft) are hot springs, spurting forth fountains from deep within the earth. The water that comes up from volcanoes is called juvenile water, which is not salty. If the Flood-stage volcanoes also continued for 40 days, they would have provided an additional source of rain, making the continuous 40-day rainfall more credible.

Now, as sole survivors of the great Flood, what are some advantages that Noah and his family enjoyed?

- No thieves, robbers, or bandits. Everything they used could be left out in the open, even overnight.

- Noah could pick the choicest land for agriculture.

- There was plenty of living space for Noah's family and for his three sons' families

- The ark, though located near the top of the mountain, provided a huge storehouse for grain and perishables and even for shelter from the heavy rain.

- All the water was clean, perfectly unpolluted.

- The ability to create huge flocks from the seven pairs of 'clean' animals.

What did we have during the Flood?
Earthquakes
Volcanic eruptions
Tectonic plate overreaches and subsidences

Torrential rains
Tidal waves

In other words, the whole surface of the earth underwent a massive reworking, so that much of the old topography was unrecognizable to Noah and his family. They undoubtedly gazed upon a brand new world, complete with an ice age.

> *The whole surface of the earth underwent a massive reworking, so that much of the old topography was unrecognizable to Noah and his family.*

All evolutionists and creationists agree on an ice age. What they do not agree on are multiple ones. To show that there was only one ice age, consider the sudden transition in climate that would have resulted at the end of the Flood: Before the Flood there were (as we suppose) subtropical conditions all over the planet. After the Flood, when the sky was emptied of its cloud canopy, the earth's surface was exposed to the direct rays of the sun during the day, but to the open, starry sky at night.; the latter meant a huge loss of heat to the atmosphere for 12 hours each day. And then the polar regions received only sunlight through a thickened atmosphere, as they do today. On top of that, since the south pole is tilted at 23 and a half degrees away from an axis perpendicular to the ecliptic, it became even more frigid than the north pole.

As the waters slowly subsided while the whole earth was exposed to a cloudless sky (although clouds were beginning to form much as they do today), the arctic and antarctic regions froze over. Though this would result in only one ice age, warming and cooling periods would not be prohibited, as meteorologists still observe and refer to them as warming or cooling cycles within an ice age. Then the seasons were felt by the ark's survivors, due to the earth's axial tilt of 23 and a half degrees. All of the other meteorological manifestations such as lightning, hail, violent storms,

Chapter 6 Aftermath of the Flood   63

tornadoes, and blizzards, came as a result of a basically unprotected sky.

The reason why just a single ice age is logical is because of the worldwide cooling effect caused by the enormous amount of ash and dust shot up by the volcanoes. This would be similar to the effect caused by Krakatoa, just one volcano, which erupted in Indonesia in 1883. The explosion and ash caused 36,417 deaths in Indonesia and countless more from wide-ranging tsunamis to other countries, the initial tidal wave registering 100 feet in height. The blast was so loud that it could be heard 2,000 miles away in Australia (the main blast being followed by several days of smaller ones). The eruptive power has been estimated to be 200 megatons of TNT, four times as powerful as the largest nuclear weapon yet detonated.

The huge quantities of ash and dust thrown into the atmosphere by this initial and smaller, follow-on eruptions darkened skies all over the world for a year, creating deep red sunsets on every continent. And most significantly, it caused the average temperature worldwide to drop by 2.2°F. If one local disruption can screen out the sun enough to cause this amount of temperature drop, how dark would the skies be if hundreds or more volcanoes

erupted from the sea within a year's time? The catch-phrase 'darkness at noon' was literally achieved within a 100-mile radius, as Krakatoa's ash plume went up 17 miles into the sky. After all the follow-on eruptions were noted by seismologists, the volcano was estimated to have blown a cubic mile of solid material into ashes.

The question always arises as to what caused the demise of the mammoths and mastodons (mastodons differing from mammoths only in the form of their molar teeth). After all the animals dispersed from the ark, it was open season for Noah and his sons to go after the wildlife. He didn't get a head start, however, because over a million mammoths had reproduced by the time of the oncoming ice age.

So man began hunting all the animals that they found edible some generations after the Flood. But bringing down a mammoth would require a joint effort, several hunters acting together; that could be one reason why the mammoths were able to reproduce so prolifically. But the main reasons for the decline of the mammoth were two-fold: eventually all the mammoths in the milder regions were hunted down for food. As the ice age moved in, the hunters unintentionally drove the mammoths toward the inhospitable arctic (the south pole being inaccessible because of ocean barriers).

Ultimately, all of the mammoths and mastodons were forced into the far north, where their coats of fur at first served them well. But when the ice age reached its peak (rather suddenly, as the mammoth remains would indicate), they all became quick-frozen because they did not want to migrate southward for fear of hunters. It is doubtful that in blizzard conditions they could retain any sense of direction at all.

From a paleontological viewpoint, the mammoths were imperious in size, 13 feet high at the shoulders. They had tusks 10 feet long, each tusk weighing 200 pounds. (An elephant tusk, for comparison, only weighs 50 pounds.) But mammals weren't the only creatures driven north and quick-frozen; often found with the mammoth remains are the woolly rhinoceros, hyena, giant stag, and other 'prehistoric' species. Many mammoths have been found in an upright position, standing on all fours in the mud. Apparently, they couldn't stop eating long enough to realize that they were

Chapter 6 Aftermath of the Flood

being put to sleep by the increasing cold wave. Six Beresoka, Russia mammoths were found with partially digested plant remains in their stomach. In Siberia alone, 50,000 mammoths have been found, yet by no means is the supply expected to diminish.

Returning to the issue of where all the water went involves a greater amount today of surface water (74%) than before the Flood (as we suppose). Fresh water diatoms have been found in several places <u>below</u> the ocean floor, indicating that parts of today's ocean are now own top of what were once fresh water seas. Biblically speaking, Psalm 104:6–9 describes the uplift of new mountains and the further deepening of pre-Flood valleys. It is assumed that there were also continental plates shifting about, since geologists have come across many cases where younger sedimentary layers are found wedged below hardened igneous rock.

Meanwhile, what happened under the seas during the slightly more than a year that the Flood waters prevailed?

- oil deposits were formed
- coal seams were pressed into existence[13]
- fossiliferous sediments were preserved in stone for antiquity

As a point of curiosity, note that the Euphrates River was present both before and after the Flood (Gen. 2:14), showing that apart from the young mountains upheaval, the old topography of the earth must not have changed that much after all. However, glacial residues, lava rock, and coral deposits contribute evidence that portions of the Atlantic Ridge were once above surface. This reinforces the premise that oceans were smaller before the Flood.

---

13 In Pennsylvania, there are 2,000 vertical feet of coal deposits following in 23 different successions.

Chapter 7
A Miscellany of So-Called Problems with the Genesis Literal Interpretation

1. Uniformitarianism Clashes with Genesis Chapters 1, 6, 7, 8

The Flood events literally moved mountains to achieve the world's topography we see today.

Uniformitarianism is the theory that all of the sedimentary strata (sandstone, limestone, shale, etc.) we see today was laid down by slow deposition over hundreds of thousands and even millions of years of geologic time. But the Genesis Flood managed to achieve the same results by compressing all of the formative geologic actions into one year of time, and all of the animal

Chapter 7 A Miscellany of So-Called Problems   67

life into six days of 'Creation.' The Flood events literally moved mountains to achieve the world's topography we see today. The final days of the Flood saw the 'young' mountains heaved up and the ocean basins collapsed to give us our mostly water-covered planet that current maps and globes depict. There are no hardened sedimentary layers that cannot be explained by the subsiding of the great Flood waters.

There are no hardened sedimentary layers that cannot be explained by the subsiding of the great Flood waters.

The most concrete example of the above opinions occurs with the eruption of Mt. St. Helens, an 8,365-foot volcanic peak in southwest Washington state. This latest eruption puts to rest the theory that it takes millions of years to convert plastic sediment into hardened rock. Mt. St. Helens erupted in 1980 and sent steam, ash, and volcanic lava high into the air, whose column was visible as far away as Seattle. It also ripped through its own crust, spewing lava, but mostly copious amounts of mud down the eastern side of the mountain. The mud, traveling initially at 500 miles an hour, was of a special kind, called pyroclastic mud, because it had bits of hardened volcanic rock mixed with it. Geologists of both the evolutionist tinge and of the creationist tinge made exhaustive studies of the mountain after the eruption settled.

Thirty-five years later, park rangers were surprised to note that the mud flow had all hardened into rock. Originally, the evolutionists did not care about the mud deposits because they had just assumed that the mud would only advance into the form of cracked, brittle, dried-out mud for their lifetime and well beyond. But it was the unbelievably thick mud flows that did the greatest damage. Not only did Mt St. Helens erupt upward, but she sent voluminous rivers of mud down the east side of the mountain. The mud was so thick that it stripped whole forests of their trees. It was the great mud flows that buried the human casualties.

As stated in our first book, there is no way to determine the age of 'prehistoric' sediment layers except by the fossil remains preserved in them. But who can determine the age of the fossils themselves except by pure guesswork? And who can determine the age of the non-fossil bearing sediments (which are far and away the moat numerous) without some basis of fact, such as the reality of the great Flood?

2. Variant Coloration of Human Skin, Hair, and Eyes

If all the humans living today descended from Adam and Eve, where did the different shades of skin originate, and why are there different color eyes and hair? Today only four main races are recognized:

                Caucasian   Negroid
                Mongoloid   Australoid

All human beings are cross-fertile, showing that they did not evolve separately from cavemen in different parts of the world. But the basic question is easily answered by taking the case of the dog. The dog (*Canus familiaris*) is considered to have evolved from the wolf, as affirmed by zoologists and evolutionists universally. After 400 years of selective breeding, we have 231 dog breeds affirmed, from the diminutive chihuahua to the bulky Alaskan malamute. From the grey wolf as a beginning, we have all kinds of fur coloration, from pure white to mottled brown to jet black. And of course we have no end to the various builds of the dog, but especially in the head and jaw. How could all these variants arise from a single pair of grey wolves?

The answer is, of course, that the genes for such variety were built in from the beginning. The original wolf had all the potential for variant coat colors, shapes, and even dispositions that can be found some place all over the world.

What about variant hair color in human beings? If we admit, as in the preceding paragraph, that different hair color genes were resident in the initial offspring of Noah, what about the suspicious geographical distribution of certain hair colors, such as black-haired humans found in the hotter climates and their blond-headed brothers found originally only in northern Europe?

Chapter 7 A Miscellany of So-Called Problems   69

Our answer involves the question of survival— dark skin and dark hair go together, where the skin is protected from the sun's ultra-violet rays by a profusion of melanin appearing on the skin's surface; without this protection, skin cancers (such as melanoma) could develop, wiping out the lighter-skinned dwellers in African and Asian equatorial areas. Conversely, dark-skinned humans who originally lived in the north would find their skin essentially blocked from absorption of solar doses of vitamin D, leaving them susceptible to malnutrition and disease.

Today, however, such changes in either direction do not lend themselves to easy measurement since the harmful or benevolent effects cannot be noticed in just one generation, but must be compounded over several generations to become observable.

What about different eye colors — blue, brown, grey, green, black, etc.? Once again, why are only the northern Europeans blue-eyed, while the rest of the world is essentially brown-eyed?

The answer could be in migration after the Flood: The brown-eyed inhabitants migrated east and south; the blue-eyed ones to the west and north. This migration could have occurred very early in the dispersion process. But as time went on, apparently the brown-eyed tribes pushed into the areas occupied by the blue-eyed ones, driving them ever northward or toward the sea, until only the colder, undesirable areas were left to the blue-eyed tribes.

3. **Does the Modern Moon with its 'Ancient' Craters Support Evolution?**

a. Astronomers tell us that the moon is slipping away from the earth by 4 inches per year. If evolution were to be believed, requiring billions of years of astrophysical history, the moon would have long since broken free of its orbit around the earth. It would only have taken tens of thousands of years, not millions or billions.

b. If the moon had been in existence for untold millions of years, the astronauts would have encountered a layer of meteoric dust so thick that they would have sunk up to their helmet in it.

70   A Reasonable Alternative to Darwin's *Origin of Species*

c. We have been led to believe that the solar system is billions of years old; why haven't the volcanoes on Earth, Venus, and Io (a moon of Jupiter) spent their upheaving activity and settled permanently, as they have on Mercury, Mars, and our moon?

4. Solar System Anomalies: Do They Support or Contradict Evolution?

How can this author say that the solar system was created in one day when it has all the earmarks of being formed over millions and billions of years? First of all, our position is that the solar system was never created in one day, as some interpreters read into Gen. 1:1. In our commentary on the first chapter of Genesis, we showed rather that God selected a planet from out of the endless trillions that he had made and began his so-called 'creation' activity in Gen.1:3 to 1:31, which then lasted for six 24-hour days.

But evolutionists have fatal flaws in their insistence on a nebular condensation theory, wherein an orbiting cloud of gas and dust supposedly circulated the sun in the same ecliptic as the sun's equatorial rotation, eventually 'condensing' and forming the (now) eight planets as we know them. Yet upon examination, each planet is found to be distinctively unique and defies the idea of formation from a homogenous nebular cloud ringed around the sun. First of all,

a. Of the eight planets, <u>Mercury</u>, though smallest, has the highest density of all the planetary bodies, thought to have an

Chapter 7 A Miscellany of So-Called Problems   71

iron core. Because of this, Mercury has a strong magnetic field, which is blatantly contrary to evolutionist thought. On any planetary body's history, the magnetic field gradually deteriorates, until there is none left; this total deterioration takes no more than a few hundred thousand years. Yet evolutionists say that Mercury (and the whole solar system) has been around for multiplied millions and even billions of years. Contrariwise, Mercury must therefore be quite young, cosmologically speaking, to display the strongest magnetic field in the solar system.

b. Then we have the sulfurous miasma that is Venus. Once thought to be a delectable sister planet of earth, Venus has been exposed as having the ultra-greenhouse effect — covered by thick clouds, Venus has a surface temperature of 900°F; if that isn't enough, the atmosphere is suffused with sulfuric acid droplets — we would last only a matter of minutes on Venus, enduring excruciating pain on the way out. Why Venus is so out of step with the other small planets does not have a good chance of being determined.

c. Mars is more like Earth than any other planet. In fact, the surface area of Mars is roughly equal to the land area of earth. But don't think of enjoying a walk on Mars without a spacesuit: the atmosphere is only 1.7 percent oxygen (vs. 20 percent on earth), diminutive enough to cause quick asphyxiation. And the air temperature at its best (at the equator, in the peak of summer), is 17 degrees below zero. Knowing that much, just imagine trying to visit Mars in its winter!

According to scientists' analysis of Mars' surface, the planet once had water, channels, and valleys. Yet now, what water that still exists lies frozen at the poles. Perhaps there is a thin twilight zone where some of the ice tries to melt at noon each day, heat being supplied by underground lava flows. But who would want to live near a thick, impermeable ice cap? For right now, Mars can be written off as one vast desert with no sufferable temperature to support plant growth even if it could draw water from the poles.

Jupiter, Saturn, Uranus, and Neptune are the gaseous giants, totally unlike the first four planets of our solar system. But each one of these giants still has its own uniqueness.

d. In the case of Jupiter, it is Jupiter's moons that make it so exceptional. The four largest satellites are called "Galilean moons," because they were first discovered with Galileo's telescope. Io, the innermost of these four moons, has over 60 active volcanoes, yet Io is only slightly smaller than our own moon, which has none. With supposedly a billion years of history for Io, all of its volcanic activity should have been spent eons ago.

e. Saturn, of course, has its rings and was until recently considered unique because of them, but Uranus and Neptune also have rings, albeit very thin ones not detectable from a telescope on earth. Yet Saturn is the only planet having a moon with an atmosphere. For as far out as it is, Titan manifests an observable atmosphere, even if it is only composed of hazy methane.

f. Uranus is out of step with the other planets because its axis of rotation is 90 degrees (instead of 0 degrees or the earth's 23½ degrees). This means that its north pole faces the sun continually, almost as if a wandering heavenly body had come by and hit Uranus, knocking the gas giant on its side. Considering the inconceivable vastness of interplanetary space, what are the odds of a wandering planetoid even finding Uranus?

g. Neptune's surface has gigantic storms as does Jupiter, strangely, for being 30 times as far from the sun as our earth, they are considered to be the most violent storms in the solar system. According to evolutionist speculation, Neptune should be a cold, dead planet. But its gigantic storms (each of which could swallow up the earth) are similar to Jupiter's huge red spot, which waxes and wanes.

h. What about Pluto, which has recently been demoted from the ninth planet to a planetoid of which the conjectured Kuiper belt has many, of sizes even bigger than Pluto itself. But Pluto has five moons of its own, one of which Charon (not

Chapter 7 A Miscellany of So-Called Problems   73

pronounced "Karon" as in Greek mythology, but "Sharon") is half the size of its master planet, causing it to rotate around Pluto in a resonance, the pair having a center of gravity somewhere in empty space between the two.

5. **The Spiral Form of Galaxies**

In spite of its mostly empty space, the universe is filled with galaxies, those seemingly floating islands of billions of stars. The farther we look into the universe, as more precise optical instruments are developed, the more we see no end to the distribution of galaxies, even no end to the so-called "clumps" of galaxies, which to our eyes appear to be in some kind of grouping. Yet in their most common form there are always more galaxies waiting for us to discover just beyond the range of the most precise telescope.

There, say the evolutionists, doesn't that verify the big bang theory?; doesn't it look as if all the galaxies were shot out at once from a gigantic explosion? It might if we could just find a central point from which the big bang occurred. All the galaxies seem to be fleeing away from us and one another, as if they were on overlapping layers if an ever-expanding balloon.

The above might be credible if it weren't for the fact that some of the galaxies are actually moving toward us, and others are moving at 45 degrees to the side of our own galactic plane. Why should we therefore doubt the big bang theory? Because the 'debris' from the big bang is not uniformly distributed in space. There are holes, and there are clumps of galaxies, yet they seem to go on forever in random distribution.

But our main objection to the big bang theory is the spiral rotation of the galaxies. True, there are some globular clusters and many isolated stars everywhere we look, but in the main they exhibit the pinwheel shape we are used to seeing for a galaxy. Because of this unique formation, we can only conclude that the universe is not being populated as the big bang would dictate.

The spirality of some galaxies makes them appear young, newly formed; others, in a perfect pinwheel shape, appear middle-aged, whole the globular clusters may represent the 'condensing' of a galaxy into its ultimate fate, a ball. Thus, the fact that the galactic formations are not all of the same shape leads us to believe that perhaps they were not exploded into infinity all at the same time.

6. Where Did Black Holes Come From?

What about the 'black holes' at the centers of some galaxies and elsewhere? Doesn't that show a timeless evolution? Don't the black holes act as a monstrous center of gravity that keeps the galaxy in a spiral rotation? And what happens when we someday arrive in a spaceship and get sucked in past the 'event horizon' of the black hole?

The black hole is located in a region of space where no material, light, or signal can escape. Within the region, the gravitational force has maximized to let nothing get away. This concept postulates that the escape velocity required at the event horizon would have to be greater than the speed of light. Einstein's general theory

Chapter 7 A Miscellany of So-Called Problems

of relativity applied to this case results from matter being squeezed (such as in a collapsed star) until it is compressed beyond what is called the Schwarzschild radius, a density of no return. In theory, any matter that happens to fall into the black hole could never come out again, but the confiscated matter would result in a proportional enlarging of the Schwarzschild radius.

Black holes supposedly start with a star much more massive than our sun. After untold eons when the star has spent its radiated energy, it suddenly collapses into either a white dwarf, neutron star, or a double-collapsed neutron star which has an 'infinite' density known as a point of singularity. Even before this final point, the compressed star continues to exist, as it shrinks the Schwarzschild radius, becoming a black hole, disappearing forever from view.

But the black hole theory comes with more questions than satisfaction. Why do some galaxies have massive black holes at their center and others (the majority) do not? How can the jets fueled by the black holes shoot out on a daily basis when the jets themselves are light-years in length? Why do some stars collapse and become neutron stars (supposedly the first stage of a black hole)? The answer to this mystery is that there probably aren't any such things as black holes. The black hole is a construct (no one has ever seen or measured it) to explain galactic flare-ups from a center of gravitational strength. Along with the black hole, hopeful astronomers have invented the 'dark matter' and 'dark energy' concepts to explain why the spiral galaxies hold together.

At this point in our astronomical knowledge (CE 2016), the black holes, dark energy, and dark matter are all theoretical, no matter how many astronomy buffs pick up on the current notions.

At this point in our astronomical knowledge (CE 2016), the black holes, dark energy, and dark matter are all <u>theoretical</u>, no matter how many astronomy buffs pick up on the current notions; their existence is just as mysterious as the quasars (see Volume I of this missive), concerning which even otherwise level-headed astronomers have gone off the deep end.

7. What Caused the Demise of Dinosaurs?

We speculated in Chapter 2 about the fate of the dinosaurs, but the question remains: What caused the demise of all the dinosaurs, from the incredible hulking monsters down to the quick-sprinting bird-size ones? The answer is not simple, yet believable. Like all the other animals, the dinosaurs made it through the Flood, with only two survivors of each species, a perfect male and female to repopulate the earth.

Now the mentality of pre-Flood humans, as revealed in legends of almost all present-day cultures is that dinosaurs were 'dragons' that needed to be exterminated. The human race fell short of its chance the first time, but the memory of daunting dinosaurs that they carried forth with them after the Flood was that the quicker the dinosaurs were eliminated, the better.

So we can reasonably suppose that the post-Flood humans didn't waste time in stalking these behemoths. The dinosaurs that the humans couldn't kill by overwhelming them with sheer numbers were hunted down by human ingenuity: all one had to do to eliminate a giant dinosaur species was to steal or smash the oversized eggs of that species. If this was kept up consistently enough, the giant species would eventually disappear. Even the vegetarian giants could be killed by gangs of humans spearing them from all angles, no doubt their carcasses serving our early ancestors as meat for many days.

What about the smaller, more agile theropods, who could outrun the humans? A quick supposition is that the early post-Flood humans who began pioneering in all directions would overreact and assume that the smaller, bird-hipped dinosaurs were really young offspring who would eventually grow to threatening size;

Chapter 7 A Miscellany of So-Called Problems   77

the smaller species could be identified by us today, but to the post-Flood humans, the only good dinosaur of any size was a dead one.

A similar question occurs regarding the flying reptiles, with the giant, leathery wingspans. Enterprising daredevil rock-climbing humans would brave the breeding cliffs to toss these reptile eggs into the sea, until the latter species dwindled to nothing.

But a more difficult question occurs with the sea-going reptiles: Did they survive the Flood? — possibly not — because as sea-going reptiles they needed to come up periodically for air; the surface of the ever-ballooning volume of water would contain almost nothing live to eat — so that they also became victims of the great Flood. What if a few did survive? Wouldn't man in his paranoia about dinosaurs try to spear these remnants from their ships? Wouldn't the few reptiles who escaped inspire all the sightings and stories and about sea monsters? And wouldn't those few remaining land-dwelling dinosaurs provide the basis for tales like "St. George and the dragon?" and for a universal dragon mythology?

So there is our conclusion: Couldn't man, if he were determined enough, wipe out today any species that he set his mind to, such as the now-endangered tigers, rhinoceri, elephants, and any of the most fierce animal predators? In fact, the endangered species list becomes lengthier every year.

8. Restriction of Marsupials Basically to the Continent of Australia

Like all marsupials, the kangaroo bears its young alive into an abdominal pouch. Why does it release its incompletely developed offspring this way? Because within the pouch are several nipples that allow the mother kangaroo to go about her business while nursing her young to maturity. So instead of fully developing her young within the womb, like placental mammals, she only partially develops her young inside her body and then lets the young fully develop outside of her belly. There can be as many young being nursed as there are nipples, so the pouch gets crowded at times.

Now, why are these marsupials found only on the continent of Australia? Simply because there were no natural predators of the kangaroo on the land down under. The path that they took to Australia, as both sides of the issue agree, was via the Malay Peninsula, because it is assumed that there was a land bridge in prehistoric times across the Malay Strait.

Well then, say the Darwinists, why are there no kangaroo fossils found in Malaysia? Our answer is also a question: The Bible mentions lions as commonplace in ancient Israel as recently as 1,000 BC.; yet why haven't we found lion fossils in modern-day Israel? Because the bones quickly deteriorate when left exposed on the open ground. Only a quick burial will preserve fossil bones, be they lion or kangaroo.

9. 'Evolution' of the Whale?

Paleontologists were quick to note that the whale, the largest animal that ever lived, be it mammal or reptile, in spite of its sea-going habitation, has mammalian characteristics: It is air-breathing, meaning that it could not stay indefinitely under water, but has to come up periodically for air, even after plunging into one of its extremely benthic dives; it nurses it young through special watertight nipples beneath the ocean surface; it is also warm-blooded (characteristic, of course, of all mammals), in spite of its oceanic environment. How then could we fail to arrive at the conclusion, as do the evolutionists, that the whale once lived on land?

Our answer is that the whale was originally made just the way it is: sea-going, and never did live on land. How can we be so sure?: simply because the sharks, dolphins, manatees, and all other airbreathing aquatic creatures would have had long evolutionist stories invented for them also.

But what about the so-called 'pelvic bone,' found toward the tail end of the whale's anatomy — doesn't that show that the whale once had legs attached and walked on land? Yet the pelvic bone is so miniscule that it could support nothing of the whale's weight; furthermore, it is a 'floating' bone, not linked by cartilage to any other bones of the whale's structure. The sperm whale, with its massive head and unbalanced weight, would have needed a special set of tyrannosaurus legs and then some, to even think about supporting its back end, much less its massive head portion, causing most of the weight to be located in the front end of the creature; this would result in legs of any size being useless for the whale.

10. The Tower of Babel

The tower of Babel is not generally understood. Why would God want to confound the tongues of the tower's builders so that they would no longer be able to understand each other? The intent of these Flood survivors was to keep communication within one known geographical area. Yet God had told their ancestor Noah, "Be fruitful, and multiply, and replenish the earth" (Gen. 9:1). In addition, God had already given Adam the same marching orders, except that he added, "and subdue it." The tower, if finished, would have been a rallying point to where the children of the Flood survivors could always come back to find their headquarters and origin. But God had a big, wide earth waiting to be filled, and no one would ever be able to stand in awe in front of the earth's natural wonders if the human populace did not explore and expand. So, after the confusion, the people gradually gathered with those they could understand and so became the nations, forming in embryo.

11. How Could Joshua Order a Long Day?

The Biblical book of Joshua, Chapter 10, verses 12-14, recounts the story of Joshua commanding the sun to stand still so that he could finish wiping out his enemies before they disappeared into the night. The sun, and the moon also, stood still for about a whole day (ca 24 hours). The only way for this to happen, say the evolutionists, is for the earth to cease rotating on its axis for 24 hours. They say, further, if that were to suddenly happen, all the loose things on the earth's surface (including human beings) would suddenly go flying eastward at 1,000 miles per hour (as measured at the equator).

Our answer is not intended to be evasive, but it is: If the earth suddenly lost its rotation, the sun would indeed appear to stand still, but the moon would keep on revolving around the earth, refuting the Bible's claim that the moon stood still "in the valley of Ajalon." So until we have a better explanation for this incident, we will just have to accept the amazing incident as one of God's special miracles.

12. How do We Account for the Unearthing of Cavemen Fossil Skulls?

The fact that all races and human beings today are cross-fertile shows that they emanated from a single pair of common ancestors. All of the different cave man types, i.e., Neanderthal[14], Cro-Magnon, etc., did not after all evolve from separate simian strains in different places of the earth.

The so-called cavemen fossil skulls fall into two categories:

1. Those skulls that have both human and apelike features but could be relegated more easily to the ape or chimpanzee species.

2. Those skulls that have both human and apelike features but appear to be more closely related to humans than apes.

Our answer is that the skulls favoring human ancestors could be clothed with flesh, suited with modern clothing, and perfectly matched to some living human being somewhere on the earth today. All over the globe, but especially in third-world countries, are living cavemen as far as facial features go. And those skulls more resembling apes or chimpanzees could find a full match easily somewhere in the family of the apes. To say that a certain skull resembles an ape and a human at the same time is to ignore the presence of both the extreme types of humans and apes dwelling in an unanticipated corner of the earth today.

13. Why is the Sudden Formation of "Young Mountains" Credible?

Evolutionists lead the way on insisting that the prehistoric land area of the earth formed one large continent, as shown:

---

14 Modernized to Neandertal.

Chapter 7 A Miscellany of So-Called Problems   81

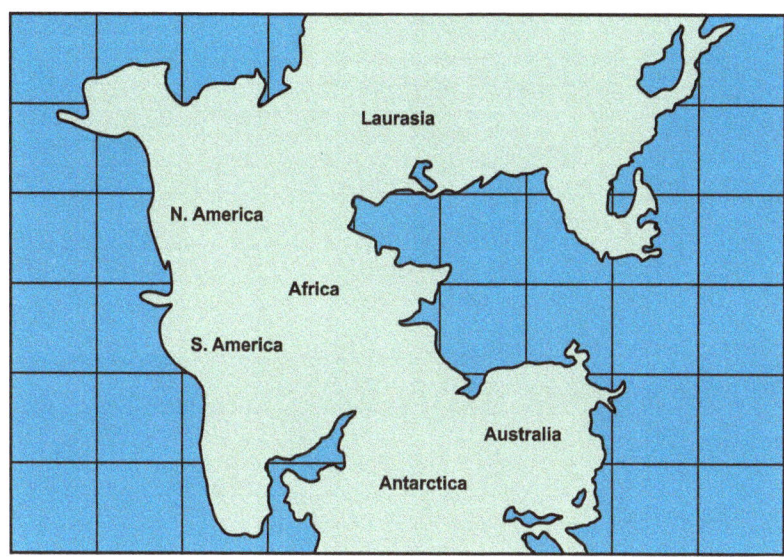

Gondwanaland

The majority of scientists accept this idea, except that evolutionists require millions of years for today's continents to have formed from Gondwanaland and Laurasia. Creationists can accept this arrangement as a pre-Flood configuration for the earth's surface. But when the Flood came and covered all, the breakup of the one-continent mass happened within a period of one year. Surely the crunching together of the tectonic plates would force new mountains up, especially when the movements occurred suddenly, geologically speaking. Otherwise, how could we account for the phrase, "and the fountains of the great deep were broken up?" (Gen.7:11).

14. How Old is the Oldest Tree?

Why is knowing what is the oldest tree on earth and its age important? Because if the tree is older than roughly 4,000 B.C. (the time of "Creation"), then a conflict exists between the Bible's chronology and that determined by today's naturalists.

The oldest known living tree is the bristlecone pine, in California's Sierra Nevada region. It has been determined to be 4,900 years old by dendrochronologists, or extant roughly from 2,900 years after Creation. Since the Flood is reckoned to have occurred at 1,656 years after Creation, we are still within chronological boundaries, and no conflict exists. This tree did not undergo extinction by reason of the Flood, as so many prehistoric species did.

15. Why are there Fresh-Water Lakes Today?

Accepting the Bible's statements that the earth was completely inundated, how do we reconcile that the subsiding water seems to have left fresh-water lakes all over the earth?

First of all, with the increased volume of water brought about by 40 days of rain, the sea became less salty than it is now. As the waters drained to form the existing oceans, the lakes that were created were less saline than the new oceans. The oceans apparently found new, exposed deposits of rock salt to dissolve and cause the waters to be heavier in their NaCl concentration.

The lakes, though they may have initially contained some concentration of salt, were subject to thousands of years of rain, which diluted the salt content even more; most of the lakes were either

Chapter 7 A Miscellany of So-Called Problems

fed by rivers or overflowed into rivers themselves. Those lakes that don't overflow will get deeper and deeper with each passing rainfall, until the salt content in that body of water becomes negligible.

16. What About the Great Intervals Required for Multiple Ice Ages on the Evolutionary Time Scale?

Supposedly, there were four major ice ages: the Nebraskan, the Illinoisan, the Kansan, and the Wisconsan (the Great), the durations of which covered tens of thousands of years. The estimated average length of each ice age was 700 years.

The reasons for a sudden ice age after the Flood were the disappearance of the cloud canopy, bringing sub-zero temperatures to the poles, the enormous amount of ash and dust shot up by the volcanoes, and the ensuing heavy, repeated snowfalls at each pole. The "Krakatoa" effect would have been worldwide, wherein the dusky skies would have caused a universal lowering of air temperature. (Interestingly, Siberia, for all its vastness, shows no signs of a prehistoric ice age.) The Pleistocene (2 million years ago) was supposed to have witnessed four ice ages and three interglacial periods. The last and largest of the ice ages (the Wisconsan) had

continental ice sheets thousands of feet thick and thousands of square miles in area.

But now for the answer: Paleontologists see four separate ice ages because they want to — there was only one great ice age. It scoured the plains of our Midwest for about 700 years and then slowly retreated northward. How do we know this? Because of the great Flood occurring 1,656 years after "Creation," there was only enough time left for a single ice age to fit into the post-Flood timetable, i.e., 700 years (all unbeknownst to Noah's descendants).

17. How Long Does it Take to Make Oil?

Consensus has it that to make oil, we only need heat, great pressure, and thousands of years, beginning with a mass of matted vegetable or animal remains. However, we have also heard about oil being produced in the laboratory under pressure from vegetation not in a matter of days, but hours.[15] Why is this significant? Because the Flood, ca 4,300 years ago, created ideal conditions to bring about the formation of oil both worldwide and in great quantities.

All the world's forests were destroyed because of the Flood. But as the leaves and other vegetation rotted, they at first formed massive floating piles of decaying matter. As a comparison, consider how much flotsam would result upon the demise of the Brazilian rainforest all in one sweep. When water-logged trees were added to the floating mats, the mats became too heavy to float and sunk to the ocean floor in unimaginably great quantities. Later, as the Flood subsided, tons of sediment were washed over the bottom-laying vegetation needed to begin the conversion process to oil. The layers of sediment, in the end sometimes miles high, provided the necessary pressure. In the case of coal, also formed during the Flood period, why are there perfect modern-looking maple leaf patterns embedded in even the deepest seems of coal?

It is not vegetation alone that produces oil. All animals have oil in their bodies (oil shale deposits are abundant with fish fossils.) Some oil deposits are derived solely from shellfish. Credible?

---

15 Reynolds, Lewis, "America the Prisoner, the Implications of Foreign Oil Addiction and Realistic Plan to End It," p. 196, Relevance Media, 2010, Charlotte, NC.

Chapter 7 A Miscellany of So-Called Problems

Interestingly, all oil wells can be viewed as floating on a pool of water. Oil and water appear together almost universally, quite reminiscent of the great Flood action.[16]

18. Why Do Coal and Oil Form Separately?

Coal is generally derived from vegetable remains, and oil generally from animal remains, although the reverse can occur. Coal seams have long been mined in Spitzbergen (above the Arctic Circle), but now they are also found in Greenland and Antarctica.

For coal to form, the vegetation must meet three conditions:

1. It must accumulate as a sizeable heap. It takes 14 feet of packed vegetation to make one foot of coal.

2. It must not be allowed to decay or mix with sedimentary deposits; it must be thoroughly excluded from air.

3. It must undergo extreme pressure as with thousands of feet of sediment or water column on top of it.

Only the great Flood could provide the conditions necessary to produce coal, which is now, ironically, found at both the north and south extremes of the earth.

19. Why are Mammoth Remains Mainly Found in the Far North?

If the antediluvian world was so globally mild, why are the mammoths and mastodons found almost exclusively just below the arctic circle? Even in the pre-Flood days there was some temperature difference depending on latitude. With their thick skin and heavy fur, these elephantine animals would reasonably have sought the coolest places they could find. When the Flood came, most of them were trapped in mud by present-day river banks. When the Flood receded, any that were trapped in mud partially above ground would have been quick-frozen by the sudden bare sky and resulting climactic change.

---

16 DiNelo, R. K and C. K. Chang, Isolation and Modification of Natural Porphyrins, New York, Academic Press, 1978

20. More Thoughts on the Gap Theory

In an attempt to account for the initial condition of the earth as described in Gen. 1:2, some theologians have proffered that there was an earlier creation with a 'gap' between theirs and ours. They envision a previous occupation of the earth by a race of humans, including all the dinosaurs, and also complications brought about by fallen angels. Even Plato mentions the drowned city of Atlantis, which rose to a height of glory, but was cursed by the gods for its unfolding wickedness.

In the case of the Genesis creation, before Adam and Eve were created, the Devil and his angelic followers had grievously offended God and were cast out of heaven (Revelation 12:4), which amounted to a loss of one-third of all the angelic beings. Once the Devil realized that he had been cursed and was evermore confined to earth, he went on a rampage and destroyed the then-existing human civilization, leaving the earth so desolate as to be "without form and void" (Gen. 1:2) It would be an untold number of years before the Adam-and-Eve creation would be made.

Those who promote the gap theory also use Gen. 1:28 — "Be fruitful and multiply, and replenish the earth," as if the earth had been previously depopulated. But the Hebrew word from which "replenish" is translated is *male*, which only means "to fill," and all modern translations so read.

21. More Thoughts on the Young Earth Theory

1. When was the earth created according to non-young earth theory?

   <u>Ans</u>. Sometime in eternity past. before Genesis 1:1 was written.

2. What about a "young earth" that gives the appearance of age, but is only between 10,000 and 13,000 years old?

   <u>Ans</u>. Unless God delights in tripping scientists over their own feet, we live on the very thin crust of a molten ball of magma. How long did it take this solid crust to form from a molten, sun-like sphere?: obviously, more than

Chapter 7 A Miscellany of So-Called Problems   87

>   13,000 years (remember Io and Europa from Chapter 1, and add Venus, our long-thought-to-be 'sister' planet, to the list also).

3. Why would God create the universe to appear to be 15 billion years old (still yet not a figure set in concrete) and then after all that apparent passage of time decide to prepare the earth for habitation?

    <u>Ans</u>. Theologically speaking, time means nothing to God. He is not constrained at all by the passage of time. Before the universe came into being (at which instant time actually began), God was. (But consider the fall of Satan from heaven some time before the human race was created.) In God's state there is no such thing as time or eternity. But to reduce a 15-billion time period to ten to thirteen thousand years is really stretching the concept of "appearance of age."

Chapter 8
Conclusions

We stated at the start of Chapter 1 of this book that the 'right' way to assess Genesis was to read nothing into it, but to let it stand on its own, literally, word for word. The major problem was what to do with the first sentence: Is it a finished summary of what had gone on in eternity past, before God had chosen planet Earth for human habitation, or is it meant to be an Introductory statement to all that follows?

> *The 'right' way to assess Genesis was to read nothing into it, but to let it stand on its own, literally, word for word.*

The author chooses the second alternative, mainly because of the otherwise imbalance between what transpires in verse 1 and 'creative' verses, 2 through 31. On one hand, if verse 1 were not an introduction, it would include the creation of the entire universe, with all its galaxies, nebulae, supernovae, quasars, etc. The

Chapter 8   Conclusions   89

greatness of infinity in all directions from earth which the universe seems to extend is far out of magnitude and proportion to the detail of the Chapter 1 verses that follow, the latter showing only a 'touch-up' on one of the (minor?) planets of one of the (minor?) solar systems of one of the billions of perceived galaxies (with yet tantalizingly more expected to be exposed as our telescopes increase in power).

In plain words, there is not a mention of the distant mysteries that we are also interested in, only a (controversial) grooming of one particular exo-planet for habitation. We therefore choose the 'Introductory' approach of verse 1 as the most reasonable in view of the minutiae that follows. All we have to do is insert a colon at the end of "earth" to make the verse introductory, doing no harm to the original Hebrew or the King James Version.

Thus, if God can be credited with creating the boundless universe in all its complexities and deep mysteries, why couldn't he have created at the same time the earth just the way he wanted it (to allow for our habitation) in one sweeping divine fiat? If the universe was just the way he wanted it, why couldn't the earth have been made that way also, in one fell stroke?

If God can be credited with creating the boundless universe in all its complexities and deep mysteries, why couldn't he have created at the same time the earth just the way he wanted it? If the universe was just the way he wanted it, why couldn't the earth have been made that way also, in one fell stroke?

(This understanding of verse 1 would, of course, override the "young earth" position, which the author regards as a grasping at straws to defend an untenable solution. The "Gap" theory must also be excluded from consideration because the earth, being created along with the rest of the universe, appeared from nothing just the way God wished it. [The Gap theory presupposes a prior fall from grace of part of the angelic beings which wrecked the earth and caused it to be left in the desolate condition described in verse 2 of Genesis 1.])

As a quick reexamination, The World Before the Flood was just as the evolutionists envision it — hot, humid, lush with vegetation, subtropical climates in what today are subarctic zones, and swamps everywhere (also, no disastrous tar pits found until after the Flood). The Lasting Impressions of the Flood Worldwide.

Chapter 8   Conclusions

Flood legends appear in so many diverse and geologically separated areas of the world that their reminiscences could not at all be attributed to a local flood.

The Flood Drastically and Forever Changed Things. All of the sedimentary rock that exists today was laid down by the Flood. Though igneous and metamorphic rock may have come from mantle pressure and lava flows, the sedimentation today is found in miles thick, alternating layers, some even crossed at 90 degrees to the layers beneath it. All of the fossils occur in strictly sedimentary rock because retrievable fossils must be quick-buried before weather and scavengers take their toll. Today's oceans are higher than those before the Flood, and the 'young' mountains released at the termination of the Flood have created permanent barriers to establish political boundaries of the world.

Climate has become controlled by the seasons, clear blue skies putting an icy grip on vast areas of our polar regions. Unknown climatic convulsions appeared to the world after the Flood, such as earthquakes, tectonic plate shifting, volcanoes, tidal waves, inland flooding, landslides, hurricanes, tornados, sinkholes, and mudslides. These and more vexations sooner or later appeared to the descendants of those who escaped the Flood in the ark.

Climate has become controlled by the seasons, clear blue skies putting an icy grip on vast areas of our polar regions. Unknown climatic convulsions appeared to the world after the Flood, such as earthquakes, tectonic plate shifting, volcanoes, tidal waves, inland flooding, landslides, hurricanes, tornados, sinkholes, and mudslides.

Even such commonplace things as dust storms could wreak havoc against our agricultural ventures, not to mention locust and other insect plagues. Expelled from Paradise we definitely traded a serene pre-Flood habitation for a volatile and unpredictable one. The seasons, it is true, usher in their change to relieve monotony, but they also have their Pandora's Box of woes. These we assume were unknown before the Flood.

We were promised, however, that such a flooding would not happen again (Gen. 9:13), the rainbow being the sign of the guarantee. [For those who are Biblically oriented, the whole earth is reserved for another violent destruction, this time by fire (II Pet. 3:10), not brought about by man's seemingly inescapable bent toward nuclear self-destruction, but by the supernatural imposition of God.]

As a final observation as to why the Bible is more plausible than evolution, consider the example of the periodic doubling of the earth's population on an average of every 40 years. This rapid increase is, of course, partially offset by the human race's incessant wars and inflicted plagues, wiping out whole ethnic populations in a generation. As a counter to this, consider the case of cavemen, which evolutionists so ardently support. Cavemen were supposed to have existed in recognizably modern appearance as early as 100,000 years ago. If we start with only two cavemen (there would have been considerably more), it would take only 36 generations to bring the earth to a condition of standing room only.

The foremost of our conclusions is that a true, literal rending of the Bible is our most reasonable approach and does not contradict true science, only the far-fetched surmisings of what well known secular scientists wish them to be.

Literal rending of the Bible is our most reasonable approach and does not contradict true science.

We invite you to view the complete
selection of titles we publish at:

www.ASPECTBooks.com

scan with your mobile
device to go directly
to our website

Please write or email us your praises, reactions, or
thoughts about this or any other book we publish at:

11 Quartermaster Circle
Fort Oglethorpe, GA 30742

Info@ASPECTBooks.com

TEACH Services, Inc., titles may be purchased in bulk for
educational, business, fund-raising, or sales promotional use.
For information, please e-mail:

BulkSales@ASPECTBooks.com

Finally if you are interested in seeing
your own book in print, please contact us at

publishing@ASPECTBooks.com

We would be happy to review your manuscript for free.